Station Amusements in New Zealand

By
Lady Barker

Edition : Culturea (culurea.fr), 34 Hérault
Contact : infos@culturea.fr
Impression : BOD, Norderstedt (Allemagne)
ISBN : 9791041984701
Date de publication : juillet 2023
Mise en page et maquettage : https://reedsy.com/
Cet ouvrage a été composé avec la police Bauer Bodoni

Station Amusements In New Zealand

Chapter I: A Bush picnic.

Since my return to England, two years ago, I have been frequently asked by my friends and acquaintances, "How did you amuse yourself up at the station?" I am generally tempted to reply, "We were all too busy to need amusement;" but when I come to think the matter over calmly and dispassionately, I find that a great many of our occupations may be classed under the head of play rather than work. But that would hardly give a fair idea of our lives there, either. It would be more correct to say perhaps, that most of our simple pleasures were composed of a solid layer of usefulness underneath the froth of fun and frolic. I purpose therefore in these sketches to describe some of the pursuits which afforded us a keen enjoyment at the time,—an enjoyment arising from perfect health, simple tastes, and an exquisite climate.

It will be as well to begin with the description of one of the picnics, which were favourite amusements in our home, nestled in a valley of the Malvern Hills of Canterbury. These hills are of a very respectable height, and constitute in fact the lowest slopes of the great Southern Alps, which rise to snow-clad peaks behind them. Our little wooden homestead stood at the head of a sunny, sheltered valley, and around it we could see the hills gradually rolling into downs, which in their turn were smoothed out, some ten or twelve miles off, into the dead level of the plains. The only drawback to the picturesque beauty of these lower ranges is the absence of forest, or as it is called there, bush. Behind the Malvern Hills, where they begin to rise into steeper ascents, lies many and many a mile of bush-clad mountain, making deep blue shadows when the setting sun brings the grand Alpine range into sharp white outline against the background of dazzling Italian sky. But just here, where my beloved antipodean home stood, we had no trees whatever, except those which we had planted ourselves, and whose growth we watched with eager interest. I dwell a little upon this point, to try to convey to any one who may glance at these pages, how we all,—dwellers among tree-less hills as we were,—longed and pined for the sights and sounds of a "bush."

Quite out of view from the house or garden, and about seven miles away, lay a mountain pass, or saddle, over a range, which was densely wooded, and from whose highest peak we could see a wide extent of timbered country. Often in our evening rides we have gone round by that saddle, in spite of a break-neck track and quicksands and bogs, just to satisfy our constant longing for green leaves, waving branches, and the twitter of birds. Whenever any wood was wanted for building a stockyard, or slabbing a well, or making a post-and-rail fence around a new paddock, we were obliged to take out a Government license to cut wood in this splendid bush. Armed with the necessary document the next step was to engage "bushmen," or woodcutters by profession, who felled and cut the timber into the proper lengths, and stacked it neatly in a clearing, where it could get dry and seasoned. These stacks were often placed in such inaccessible and rocky parts of the steep mountain side, that they had to be brought down to the flat in rude little sledges, drawn by a bullock, who required to be trained to the work, and to possess so steady and equable a disposition as to be indifferent to the annoyance of great logs of heavy wood dangling and bumping against his heels as the sledge pursued its uneven way down the bed of a mountain torrent, in default of a better road.

Imagine, then, a beautiful day in our early New Zealand autumn. For a week past, a furious north-westerly gale had been blowing down the gorges of the Rakaia and the Selwyn, as if it had come out of a funnel, and sweeping across the great shelterless plains with irresistible force. We had been close prisoners to the house all those days, dreading to open a door to go out for wood or water, lest a terrific blast should rush in and whip the light shingle roof off. Not an animal could be seen out of doors; they had all taken shelter on the lee-side of the gorse hedges, which are always planted round a garden to give the vegetables a chance of coming up. On the sky-line of the hills could be perceived towards evening, mobs of sheep feeding with their heads up-wind, and travelling to the high camping-grounds which they always select in preference to a valley. The yellow tussocks were bending all one way, perfectly flat to the ground, and the shingle on the gravel walk outside rattled like hail against the low latticed windows. The uproar from the gale was indescribable, and the little fragile house swayed and shook as the furious gusts hurled themselves against it. Inside its shelter, the pictures were blowing out from the walls, until I expected them to be shaken off their hooks even in those rooms which had plank walls lined with papered canvas; whilst in the kitchen, store-room, etc., whose sides were made of cob, the dust blew in fine clouds from the pulverized walls, penetrating even to the dairy, and settling half an inch thick on my precious cream. At last, when our skin felt like tightly drawn parchment, and our ears and eyes had long been filled with powdered earth, the wind dropped at sunset as suddenly as it had risen five days before. We ventured out to breathe the dust-laden atmosphere, and to look if the swollen creeks (swollen because snow-fed) had done or threatened to do any mischief, and saw on the south-west horizon great fleecy masses of cloud driving rapidly up before a chill icy breeze. Hurrah, here comes a sou'-wester! The parched-up earth, the shrivelled leaves, the dusty grass, all needed the blessed damp air. In an hour it was upon us. We had barely time to house the cows and horses, to feed the fowls, and secure them in their own shed, and to light a roaring coal (or rather lignite, for it is not true coal) fire in the drawing-room, when, with a few warning splashes, the deluge of cold rain came steadily down, and we went to sleep to the welcome sound of its refreshing patter.

All that I have been describing was the weather of the past week. Disagreeable as it might have been, it was needed in both its hot and cold, dry and wet extremes, to make a true New Zealand day. The furious nor'-wester had blown every fleck of cloud below the horizon, and dried the air until it was as light as ether. The "s'utherly buster," on the other hand, had cooled and refreshed everything in the most delicious way, and a perfect day had come at last. What words can describe the pleasure it is to inhale such an atmosphere? One feels as if old age or sickness or even sorrow, could hardly exist beneath such a spotless vault of blue as stretched out above our happy heads. I have often been told that this feeling of intense pleasure on a fine day, which is peculiar to New Zealand, is really a very low form of animal enjoyment. It may be so, but I only know that I never stood in the verandah early in the morning of such a day as I am trying to sketch in pen and ink now, without feeling the highest spiritual joy, the deepest thankfulness to the loving Father who had made His beautiful world so fair, and who would fain lead us through its paths of pleasantness to a still more glorious, home, which will be free from the shadows brooding from beneath sin's out-stretched wings over this one. As I stood in the porch I have often fancied I could seethe animals and even the poultry expressing in dumb brute fashion, their joy and gratitude to the God from whom all blessings flow.

But to return to the verandah, although we have never left it. Presently F— — came out, and I said with a sigh, born of deep content and happiness, "What a day!" "Yes," answered F— —: "a heavenly day indeed: well worth waiting for. I want to go and see how the men are getting on in the bush. Will you like to come too?" "Of course I will. What can be more enchanting than the prospect of spending such sunny hours in that glorious bush?" So after breakfast I give my few simple orders to the cook, and prepare, to pack a "Maori kit," or flat basket made of flax, which could be fastened to my side-saddle, with the preparations for our luncheon. First some mutton chops had to be trimmed and prepared, all ready to be cooked when we got there. These were neatly folded up in clean paper; and a little packet of tea, a few lumps of white sugar, a tiny wooden contrivance for holding salt and pepper, and a couple of knives and forks, were added to the parcel.

So much for the contents of the basket. They needed to be carefully packed so as not to rattle in any way, or Helen, my pretty bay mare, would soon have got rid of the luncheon—and me. I wrapped up three or four large raw potatoes in separate bits of paper, and slipped them into F— —'s pockets when he was looking another way, and then began the real difficulty of my picnic: how was the little tin tea-pot and an odd delf cup to be carried? F— — objected to put them also in his pocket, assuring me that I could make very good tea by putting my packet of the fragrant leaves into the bushmen's kettle, and drinking it afterwards out of one of their pannikins. He tried to bribe me to this latter piece of simplicity by promising to wash the tin pannikin out for me first. Now I was not dainty or over particular; I could not have enjoyed my New Zealand life so thoroughly if I had been either; but I did not like the idea of using the bushmen's tea equipage. In the first place, the tea never tastes the same when made in their way, and allowed to boil for a moment or two after the leaves have been thrown in, before the kettle is taken off the fire; and in the next place, it is very difficult to drink tea out of a pannikin; for it becomes so hot directly we put the scalding liquid into it, that long after the tea is cool enough to drink, the pannikin still continues too hot to touch. But I said so pathetically, "You know how wretched I am without my tea," that F— —'s heart relented, and he managed to stow away the little teapot and the cup. That cup bore a charmed life. It accompanied me on all my excursions, escaping unbroken; and is, I believe, in existence now, spending its honoured old age in the recesses of a cupboard.

After the luncheon, the next question to be decided is, which of the dogs are to join the expedition. Hector, of course; he is the master's colley, and would no more look at a sheep, except in the way of business, than he would fly. Rose, a little short-haired terrier, was the most fascinating of dog companions, and I pleaded hard for her, as she was an especial pet; though there were too many lambs belonging to a summer lambing (in New Zealand the winter is the usual lambing season) in the sheltered paddocks beneath the bush, to make it quite safe for her to be one of the party. She would not kill or hurt a lamb on any account, but she always appeared anxious to play with the little creatures; and as her own spotless coat was as white as theirs, she often managed to get quite close to a flock of sheep before they perceived that she belonged to the dreaded race of dogs. When the timid animals found out their mistake, a regular stampede used to ensue; and it was not supposed to be good for the health of the old or young sheep to hurry up the hill-sides in such wild fashion as that in which they rushed away from Rose's attempts to intrude on their society. Nettle may come, for he is but a tiny terrier, and so fond of his mistress that he never strays a yard away from her horse's heels. Brisk, my beautiful, stupid water-spaniel, is also allowed an outing. He is perfect to look at, but not having had any educational advantages in his youth, is an utter fool; amiable, indeed, but not the less a fool. Garibaldi, another colley, is suffering a long penal sentence of being tied up to his barrel, on account of divers unlawful chases after sheep which were not wanted; and dear old

Jip, though she pretends to be very anxious to accompany us; is far too fat and too rheumatic to keep pace with our long stretching gallop up the valley.

At last we were fairly off about eleven o'clock, and an hour's easy canter, intersected by many "flat-jumps," or rather "water-jumps," across the numerous creeks, brought unto the foot of the bush-clad mountain. After that our pace became a very sober one, as the track resembled a broken rocky staircase more than a bridle-path. But such as it was, our sure-footed horses carried us safely up and down its rugged steeps, without making a single false step. No mule can be more sure-footed than a New Zealand horse. He will carry his rider anywhere, if only that rider trusts entirely to him, nor attempts to guide him in any way. During the last half-hour of our slow and cat-like climb, we could hear the ring of the bushmen's axes, and the warning shouts preceding the crashing fall of a Black Birch. Fallen logs and deep ruts made by the sledges in their descent, added to the difficulties of the track; and I was so faint-hearted as to entreat piteously, on more than one occasion, when Helen paused and shook her head preparatory to climbing over a barricade, to be "taken off." But F—— had been used to these dreadful roads for too many years to regard them in the same light as I did, and would answer carelessly, "Nonsense: you're as safe as if you were sitting in an arm-chair." All I can say is, it might have been so, but I did not feel at all like it.

However, the event proved him to have been right, and we reached the clearing in safety. Here we dismounted, and led the horses to a place where they could nibble some grass, and rest in the cool shade. The saddles and bridles were soon removed, and halters improvised out of the New Zealand flax, which can be turned to so many uses. Having provided for the comfort of our faithful animals, our next step was to look for the bushmen. The spot which we had reached was their temporary home in the heart of the forest, but their work was being carried on elsewhere. I could not have told from which side the regular ringing axe-strokes proceeded, so confusing were the echoes from the cliffs around us; but after a moment's silent pause F—— said, "If we follow that track (pointing to a slightly cleared passage among the trees) we shall come upon them." So I kilted up my linsey skirt, and hung up my little jacket, necessary for protection against the evening air, on a bough out of the wekas' reach, whilst I followed F— — through tangled creepers, "over brake, over brier," towards the place from whence the noise of falling trees proceeded. By the time we reached it, our scratched hands and faces bore traces of the thorny undergrowth which had barred our way; but all minor discomforts were forgotten in the picturesque beauty of the spot. Around us lay the forest-kings, majestic still in their overthrow, whilst substantial stacks of cut-up and split timber witnessed to the skill and industry of the stalwart figures before us, who reddened through their sunburn with surprise and shyness at seeing a lady. They need not have been afraid of me, for I had long ago made friends with them, and during the preceeding winter had established a sort of night-school in my dining-room, for all the hands employed on the station, and these two men had been amongst my most constant pupils. One of them, a big Yorkshire-man, was very backward in his "larning," and though he plodded on diligently, never got beyond the simplest words in the largest type. Small print puzzled him at once, and he had a habit of standing or sitting with his back to me whilst repeating his lessons. Nothing would induce him to face me. The moment it became his turn to go on with the chapter out of the Bible, with which we commenced our studies, that instant he turned his broad shoulders towards me, and I could only, hear the faintest murmurs issuing from the depths of a great beard. Remonstrance would have scared my shy pupil away, so I was fain to put up with his own method of instruction.

But this is a digression, and I want to make you see with my eyes the beautiful glimpses of distant country lying around the bold wooded cliff on which we were standing. The ground fell away from our feet so completely in some places, that we could see over the tops of the high trees around us, whilst in others the landscape appeared framed in an arch of quivering foliage. A noisy little creek chattered and babbled as it hurried along to join its big brother down below, and kept a fringe of exquisite ferns, which grew along its banks, brightly green by its moisture. Each tree, if taken by itself, was more like an umbrella than anything else to English eyes, for in these primitive forests, where no kind pruning hand has ever touched them, they shoot up, straight and branchless, into the free air above, where they spread a leafy crown out to the sunbeams. Beneath the dense shade of these matted branches grew a luxuriant shrubbery, whose every leaf was a marvel of delicate beauty, and ferns found here a home such as they might seek elsewhere in vain. Flowers were very rare, and I did not observe many berries, but these conditions vary in different parts of the beautiful middle island.

That was a fair and fertile land stretching out before us, intersected by the deep banks of the Rakaia, with here and there a tiny patch of emerald green and a white dot, representing the house and English grass paddock of a new settler. In the background the bush-covered mountains rose ever higher and higher in bolder outline, till they shook off their leafy clothing, and stood out in steep cliffs and scaurs from the snow-clad glacier region of the mountain range running from north to south, and forming the back bone of the island. I may perhaps make you see the yellow, river-furrowed plains, and the great confusion of rising ground behind them, but cannot make you see, still less feel, the atmosphere around, quivering in a summer haze in the valley beneath, and stirred to the faintest summer wind-sighs as it moved among the pines and birches overhead. Its lightness was its most striking peculiarity. You felt as if your lungs could never weary of inhaling deep breaths of such an air. Warm without oppression, cool without a chill. I can find nothing but paradoxes to describe it. As for fatigue, one's muscles might get tired, and need rest, but the usual depression and weariness attending over-exertion could not exist in such an atmosphere. One felt like a happy child; pleased at nothing, content to exist where existence was a pleasure.

You could not find more favourable specimens of New Zealand colonists than the two men, Trew and Domville, who stood before us in their working dress of red flannel shirts and moleskin trousers, "Cookham" boots and digger's plush hats. Three years before this day they had landed at Port Lyttleton, with no other capital than their strong, willing arms, and their sober, sensible heads. Very different is their appearance to-day from what it was on their arrival; and the change in their position and circumstances is as great. Their bodily frames have filled out and developed under the influence of the healthy climate and abundance of mutton, until they look ten years younger and twice as strong, and each man owns a cottage and twenty acres of freehold land, at which he works in spare time, as well as having more pounds than he ever possessed pence in the old country, put safely away in the bank. There can be no doubt about the future of any working man or woman in our New Zealand colonies. It rests in their own hands, under God's blessing, and the history of the whole human race shows us that He always has blessed honest labour and rightly directed efforts to do our duty in this world. Sobriety and industry are the first essentials to success. Possessing these moral qualifications, and a pair of hands, a man may rear up his children in those beautiful distant lands in ignorance of what hunger; or thirst, or grinding poverty means. Hitherto the want of places of worship, and schools for the children, have been a sad drawback to the material advantages of colonization at the Antipodes; but these blessings are increasing every day, and the need of them creates the supply.

The great mistake made in England, next to that of sending out worthless idle paupers, who have never done a hand's turn for themselves here, and are still less likely to do it elsewhere, is for parents and guardians to ship off to New Zealand young men who have received the up-bringing and education of gentlemen, without a shilling in their pockets, under the vague idea that something will turn up for them in a new place. There is nothing which can turn up, for the machinery of civilization is reduced to the most primitive scale in these countries; and I have known 500 pounds per annum regarded as a monstrous salary to be drawn by a hard-worked official of some twenty years standing and great experience in the colony. From this we may judge of the chances of remunerative employment for a raw unfledged youth, with a smattering of classical learning. At first they simply "loaf" (as it is called there) on their acquaintances and friends. At the end of six months their clothes are beginning to look shabby; they feel they ought to do something, and they make day by day the terrible discovery that there is nothing for them to do in their own rank of life.

Many a poor clergyman's son, sooner than return to the home which has been so pinched to furnish forth his passage money and outfit, takes a shepherd's billet, though he generally makes a very bad shepherd for the first year or two; or drives bullocks, or perhaps wanders vaguely over the country, looking for work, and getting food and lodging indeed, for inhospitality is unknown, but no pay. Sometimes they go to the diggings, only to find that money is as necessary there as anywhere, and that they are not fitted to dig in wet holes for eight or ten hours a day. Often these poor young men go home again, and it is the best thing they can do, for at least they have gained some knowledge of life, on its dark as well as its brighter side. But still oftener, alas, they go hopelessly to the bad, degenerating into billiard markers, piano players at dancing saloons, cattle drivers, and their friends probably lose sight of them.

Once I was riding with my husband up a lovely gulley, when we heard the crack of a stockwhip, sounding strangely through the deep eternal silence of a New Zealand valley, and a turn of the track showed us a heavy, timber-laden bullock-waggon labouring slowly along. At the head of the long team sauntered the driver, in the usual rough-and-ready costume, with his soft plush hat pulled low over his face, and pulling vigorously at a clay pipe. In spite of all the outer surroundings, something in the man's walk and dejected attitude struck my imagination, and I made some remark to my companion. The sound of my voice reached the bullock-driver's ears; he looked up, and on seeing a lady, took his pipe out of his mouth, his hat off his head, and forcing his beasts a little aside, stood at their head to let us pass. I smiled and nodded, receiving in return a perfect and profound bow, and the most melancholy glance I have ever seen in human eyes. "Good gracious, F— —," I cried, when we had passed, "who is that man?" "That is Sir So-and-So's third son," he replied: "they sent him out here without a shilling, five years ago, and that is what he has come to: a working man, living with working men. He looks heart-broken, poor fellow, doesn't he?" I, acting upon impulse, as any woman would have done, turning back and rode up to him, finding it very difficult to frame my pity and sympathy in coherent words. "No thank you, ma'am," was all the answer I could get, in the most refined, gentlemanly tone of voice: "I'm very well as I am. I should only have the struggle all over again if I made any change now. It is the truest kindness to leave me alone." He would not even shake hands with me; so I rode back; discomfited, to hear from F— — that he had made many attempts to befriend him, but without success. "In fact," concluded F— —, with some embarrassment, "he drinks dreadfully, poor fellow. Of course that is the secret of all his wretchedness, but I believe despair drove him to it in the first instance."

I have also known an ex-dragoon officer working as a clerk in an attorney's office at fifteen shillings a week, who lived like a mechanic, and yet spake and stepped like his old self; one listened involuntarily for the clink of the sabre and spur whenever he moved across the room.

This has been a terrible digression, almost a social essay in fact; but I have it so much at heart to dissuade fathers and mothers from sending their sons so far away without any certainty of employment. Capitalists, even small ones, do well in New Zealand: the labouring classes still better; but there is no place yet for the educated gentleman without money, and with hands unused to and unfit for manual labour and the downward path is just as smooth and pleasant at first there, as anywhere else.

Trew and Domville soon got over their momentary shyness, and answered my inquiries about their families. Then I had a short talk with them, but on the principle that it is "ill speaking to a fasting man," we agreed to adjourn to the clearing, where they had built a rough log hut for temporary shelter, and have our dinner. They had provided themselves with some bacon; but were very glad to accept of F— —'s offer of mutton, to be had for the trouble of fetching it. When we reached the little shanty, Trew produced some capital bread, he had baked the evening before in a camp-oven; F— —'s pockets were emptied of their load of potatoes, which were put to roast in the wood embers; rashers of bacon and mutton chops spluttered and fizzed side-by-side on a monster gridiron with tall feet, so as to allow it to stand by itself over the clear fire, and we turned our chops from time to time by means of a fork extemporized out of a pronged stick.

Over another fire, a little way to leeward, hung the bushmen's kettle on an iron tripod, and, so soon as it boiled, my little teapot was filled before

Domville threw in his great fist-full of tea. I had brought a tiny phial of cream in the pocket of my saddle, but the men thought it spoiled the flavour of the tea, which they always drink "neat," as they call it. The Temperance Society could draw many interesting statistics from the amount of hard work which is done in New Zealand on tea. Now, I am sorry to say, beer is creeping up to the stations, and is served out at shearing time and so on; but in the old days all the hard work used to be done on tea, and tea alone, the men always declaring they worked far better on it than on beer. "When we have as much good bread and mutton as we can eat," they would say, "we don't feel to miss the beer we used to drink in England;" and at the end of a year or two of tea and water-drinking, their bright eyes and splendid physical condition showed plainly enough which was the best kind of beverage to work, and work hard too, upon.

So there we sat round the fire: F— — with the men, and I, a little way off, out of the smoke, with the dogs. Overhead, the sunlight streamed down on the grass which had sprung up, as it always does in a clearing; the rustle among the lofty tree tops made a delicious murmur high up in the air; a waft of cool breeze flitted past us laden with the scent of newly-cut wood (and who does not know that nice, clean perfume?); innumerable paroquets almost brushed us with their emerald-green wings, whilst the tamer robin or the dingy but melodious bell-bird came near to watch the intruders. The sweet clear whistle of the tui or parson-bird—so called from his glossy black suit and white wattles curling exactly where a clergy-man's bands would be,—could be heard at a distance; whilst overhead the soft cooing of the wild pigeons, and the hoarse croak of the ka-ka or native parrot, made up the music of the birds' orchestra. Ah, how delicious it all was,—the Robinson Crusoe feel of the whole thing; the heavenly air, the fluttering leaves, the birds' chirrups and whistle, and the foreground of happy, healthy men!

Rose and I had enough to do, even with Nettle's assistance, in acting as police to keep off those bold thieves, the wekas, who are as impudent as they are tame and fearless. In appearance they resemble exactly a stout hen pheasant, without its long tail; but they belong to the apterix family, and have no wings, only a tiny useless pinion at each shoulder, furnished with a claw like a small fish-hook: what is the use of this claw I was never able to discover. When startled or hunted, the weka glides, for it can scarcely be called running, with incredible swiftness and in perfect silence, to the nearest cover. A tussock, a clump of flax, a tuft of tall tohi grass, all serve as hiding-places; and, wingless as she is, the weka can hold her own very well against her enemies, the dogs. I really believe the great desire of Brisk's life was to catch a weka. He started many, but used to go sniffing and barking round the flax bush where it had taken refuge at first, long after the clever, cunning bird had glided from its shelter to another cover further off.

After dinner was over and Domville had brought back the tin plates and pannikins from the creek where he had washed them up, pipes were lighted, and a few minutes smoking served to rest and refresh the men, who had been working since their six o'clock breakfast. The daylight hours were too precious however to be wasted in smoking. Trew and Domville would not have had that comfortable nest-egg standing in their name at the bank in Christchurch, if they had spent much time over their pipes; so after a very short "spell" they got up from the fallen log of wood which had served them for a bench, and suggested that F—— should accompany them back to where their work lay. "You don't mind being left?" asked F——. "Certainly not," replied I. "I have got the dogs for company, and a book in my pocket. I daresay I shall not read much, however, for it is so beautiful to sit here and watch the changing lights and shadows."

And so it was, most beautiful and thoroughly delightful. I sat on the short sweet grass, which springs upon the rich loam of fallen leaves the moment sunlight is admitted into the heart of a bush. No one plants it; probably the birds carry the seeds; yet it grows freely after a clearing has been made. Nature lays down a green sward directly on the rich virgin mould, and sets to work besides to cover up the unsightly stems and holes of the fallen timber with luxuriant tufts of a species of hart's-tongue fern, which grows almost as freely as an orchid on decayed timber. I was so still and silent that innumerable forest birds came about me. A wood pigeon alighted on a branch close by, and sat preening her radiant plumage in a bath of golden sunlight. The profound stillness was stirred now and then by a soft sighing breeze which passed over the tree tops, and made the delicate foliage of the undergrowth around me quiver and rustle. I had purposely scattered the remains of our meal in a spot where the birds could see the crumbs, and it was not long before the clever little creatures availed themselves of the unexpected feast. So perfectly tame and friendly were they, that I felt as if I were the intruder, and bound by all the laws of aerial chivalry to keep the peace. But this was no easy matter where Rose and Nettle were concerned, for when an imprudent weka appeared on the sylvan scene, looking around-as if to say, "Who's afraid?" it was more than I could do to keep the little terriers from giving chase. Brisk, too, blundered after them, but I had no fear of his destroying the charm of the day by taking even a weka's life.

Thus the delicious afternoon wore on, until it was time to boil the kettle once more, and make a cup of tea before setting out homewards. The lengthening shadows added fresh tenderness and beauty to the peaceful scene, and the sky began to paint itself in its exquisite sunset hues. It has been usual to praise the tints of tropic skies when the day is declining; but never, in any of my wanderings to East and West Indies, have I seen such gorgeous evening colours as those which glorify New Zealand skies.

A loud coo-ee summoned F— — to tea, and directly afterwards the horses were re-saddled, the now empty flax basket filled with the obnoxious teapot and cup, wrapped in many layers of flax leaves, to prevent their rattling, and we bade good night to the tired bushmen. We left them at their tea, and I was much struck to observe that though they looked like men who had done a hard day's work, there was none of the exhaustion we often see in England depicted on the labouring man's face. Instead of a hot crowded room, these bushmen were going to sleep in their log hut, where the fresh pure air could circulate through every nook and cranny. They had each their pair of red blankets, one to spread over a heap of freshly cut tussocks, which formed a delicious elastic mattrass, and the other to serve as a coverlet. During the day these blankets were always hung outside on a tree, out of the reach of the most investigating weka. You may be sure I had not come empty-handed in the way of books and papers, and my last glance as I rode away rested on Trew opening a number of Good Words with the pleased-expression of a child examining a packet of toys.

And so we rode slowly home through the delicious gloaming, with the evening air cooled to freshness so soon as the sun had sunk below the great mountains to the west, from behind which he shot up glorious rays of gold and crimson against the blue ethereal sky, causing the snowy peaks to look more exquisitely pure from the background of gorgeous colour. During the flood of sunlight all day, we had not perceived a single fleck of cloud; but now lovely pink wreaths, floating in mid-air, betrayed that here and there a "nursling of the sky" lingered behind the cloud-masses which we thought had all been blown away yesterday.

The short twilight hour was over, and the stars were filtering their soft radiance on our heads by the time we heard the welcoming barks of the homestead, and saw the glimmer of the lighted lamp in our sitting-room, shining out of the distant gloom. And so ended, in supper and a night of deep dreamless sleep, one of the many happy picnic days of my New Zealand life.

Chapter II: Eel-fishing.

One of the greatest drawbacks in an English gentleman's eyes to living in New Zealand is the want of sport. There is absolutely none. There used to be a few quails, but they are almost extinct now; and during four years' residence in very sequestered regions I only saw one. Wild ducks abound on some of the rivers, but they are becoming fewer and shyer every year. The beautiful Paradise duck is gradually retreating to those inland lakes lying at the foot of the Southern Alps, amid glaciers and boulders which serve as a barrier to keep back his ruthless foe. Even the heron, once so plentiful on the lowland rivers, is now seldom seen. As I write these lines a remorseful recollection comes back upon me of overhanging cliffs, and of a bend in a swirling river, on whose rapid current a beautiful wounded heron—its right wing shattered—drifts helplessly round and round with the eddying water, each circle bringing it nearer in-shore to our feet. I can see now its bright fearless eye, full of suffering, but yet unconquered: its slender neck proudly arched, and bearing up the small graceful head with its coronal or top-knot raised in defiance, as if to protest to the last against the cruel shot which had just been fired. I was but a spectator, having merely wandered that far to look at my eel-lines, yet I felt as guilty as though my hand had pulled the trigger. Just as the noble bird drifted to our feet,—for I could not help going down to the river's edge, where Pepper (our head shepherd) stood, looking very contrite,—it reared itself half out of the water, with a hissing noise and threatening bill, resolved to sell its liberty as dearly as it could; but the effort only spread a brighter shade of crimson on the waters surface for a brief moment, and then, with glazing eye and drooping crest, the dying creature turned over on its side and was borne helpless to our feet. By the time Pepper extended his arm and drew it in, with the quaint apology, "I'm sorry I shot yer, old feller! I, am, indeed," the heron was dead; and that happened to be the only one I ever came across during my mountain life. Once I saw some beautiful red-shanks flying down the gorge of the Selwyn, and F— — nearly broke his neck in climbing the crag from whence one of them rose in alarm at the noise of our horses' feet on the shingle. There were three eggs in the inaccessible cliff-nest, and he brought me one, which I tried in vain to hatch under a sitting duck. Betty would not admit the intruder among her own eggs, but resolutely pushed it out of her nest twenty times a day, until at last I was obliged to blow it and send it home to figure in a little boy's collection far away in Kent.

I have seen very good blue duck shooting on the Waimakiriri river, but 50 per cent. of the birds were lost for want of a retriever bold enough to face that formidable river. Wide as was the beautiful reach, on whose shore the sportsmen stood, and calmly as the deep stream seemed to glide beneath its high banks, the wounded birds, flying low on the water, had hardly dropped when they disappeared, sucked beneath by the strong current, and whirled past us in less time than it takes one to write a line. We had retrievers with us who would face the waves of an inland lake during a nor'-wester,—which is giving a dog very high praise indeed; but there was no canine Bayard at hand to brave those treacherous depths, and bring out our game, so the sport soon ceased; for what was the good of shooting the beautiful, harmless creatures when we could not make use of them as food?

I often accompanied F—— on his eel-fishing expeditions, but more for the sake of companionship than from any amusement I found in the sport. I may here confess frankly that I cannot understand anyone being an inveterate eel-fisher, for of all monotonous pursuits, it is the most self-repeating in its forms. Even the first time I went out I found it delightful only in anticipation; and this is the one midnight excursion which I shall attempt to re-produce for you.

It had been a broiling midsummer day, too hot to sit in the verandah, too hot to stroll about the garden, or go for a ride, or do anything in fact, except bask like a lizard in the warm air. New Zealand summer weather, however high the thermometer, is quite different from either tropical or English heat. It is intensely hot in the sun, but always cool in the shade. I never heard of an instance of sun-stroke from exposure to the mid-day sun, for there always was a light air—often scarcely perceptible until you were well out in the open,—to temper the fierce vertical rays. It sometimes happened that I found myself obliged, either for business or pleasure, to take a long ride in the middle of a summer's day, and my invariable reflection used to be, "It is not nearly so hot out of doors as one fancies it would be." Then there is none of the stuffiness so often an accompaniment to our brief summers, bringing lassitude and debility in its train. The only disadvantage of an unusually hot season with us was, that our already embrowned complexions took a deeper shade of bronze; but as we were all equally sun-burnt there was no one to throw critical stones.

What surprised me most was the utter absence of damp or miasma. After a blazing day, instead of hurrying in out of reach of poisonous vapours as the tropic-dweller must needs do, we could linger bare-headed, lightly clad, out of doors, listening to the distant roar of a river, or watching the exquisite tints of the evening sky. I dwell on this to explain that in almost any other country there would have been risk in remaining out at night after such still, hot days.

On this particular evening, during my first summer in the New Zealand Malvern Hills, after we had watered my pet flowers near the house, and speculated a good deal as to whether the mignonette seed had all been blown out of the ground by the last nor'-wester or not, F— — said, "I shall go eel-fishing to-night to the creek, down the flat. Why don't you come too? I am sure you would like it." Now, I am sorry to say that I am such a thorough gipsy in my tastes that any pursuit which serves as an excuse for spending hours in the open air, is full of attraction for me; consequently, I embraced the proposal with ardour, and set about gathering, under F— —'s directions, what seemed to bid fair to rival the collection of an old rag-and-bottle merchant. First of all, there was a muster of every empty tin match-box in the little house; these were to hold the bait-bits of mutton and worms. Then I was desired to hunt up all the odds and ends of worsted which lurked in the scrap-basket. A forage next took place in search of string, but as no parcels were ever delivered in that sequestered valley, twine became a precious and rare treasure. In default of any large supply being obtainable, my lamp and candle-wick material was requisitioned by F— — (who, by the way, is a perfect Uhlan for getting what he wants, when bent on a sporting expedition); and lastly, one or two empty flour-sacks were called for. You will see the use of this heterogeneous collection presently.

It was of no use starting until the twilight had darkened into a cloudy, moonless night; so, after our seven o'clock supper, we adjourned into the verandah to watch F— — make a large round ball, such as children play with, out of the scraps of worsted with which I had furnished him. Instead of cutting the wool into lengths, however, it was left in loops; and I learned that this is done to afford a firm hold for the sharp needle-like teeth of an inquisitive eel, who might be tempted to find out if this strange round thing, floating near his hole, would be good to eat. I was impatient as a child,—remember it was my first eel-fishing expedition,—and I thought nine o'clock would never come, for I had been told to go and dress at that hour; that is to say, I was to change my usual station-costume, a pretty print gown, for a short linsey skirt, strong boots and kangaroo-skin gaiters. F— —, and our cadet, Mr. U— —, soon appeared, clad in shooting coats instead of their alpaca costumes, and their trousers stuffed into enormous boots, the upper leathers of which came beyond their knees.

"Are we going into the water?" I timidly inquired.

"Oh, no,—not at all: it is on account of the Spaniards."

No doubt this sounds very unintelligible to an English reader; but every colonist who may chance to see my pages will shiver at the recollection of those vegetable defenders of an unexplored region in New Zealand. Imagine a gigantic artichoke with slender instead of broad leaves, set round in dense compact order. They vary, of course, in size, but in our part of the world four or six feet in circumference and a couple of feet high was the usual growth to which they attained, though at the back of the run they were much larger. Spaniards grow in clusters, or patches, among the tussocks on the plains, and constitute a most unpleasant feature of the vegetation of the country. Their leaves are as firm as bayonets, and taper at the point to the fineness of a needle, but are not nearly so easily broken as a needle would be. No horse will face them, preferring a jump at the cost of any exertion, to the risk of a stab from the cruel points. The least touch of this green bayonet draws blood, and a fall intoa Spaniard is a thing to be remembered all one's life. Interspersed with the Spaniards are generally clumps of "wild Irishman," a straggling sturdy bramble, ready to receive and scratch you well if you attempt to avoid the Spaniard's weapons. Especially detrimental to riding habits are wild Irishmen; and there are fragments of mine, of all sorts of materials and colours, fluttering now on their thorny branches in out-of-the-way places on our run. It is not surprising, therefore, that we guarded our legs as well as we could against these foes to flesh and blood.

"We are rather early," said the gentlemen, as I appeared, ready and eager to start; "but perhaps it is all the better to enable you to see the track." They each flung an empty sack over their shoulders, felt in their pockets to ascertain whether the matches, hooks, boxes of bait, etc., were all there, and then we set forth.

At first it appeared as if we had stepped from the brightness of the drawing-room into utter and pitchy blackness; but after we had groped for a few steps down the familiar garden path, our eyes became accustomed to the subdued light of the soft summer night. Although heavy banks of cloud,—the general precursors of wind,—were moving slowly between us and the heavens, the stars shone down through their rifts, and on the western horizon a faint yellowish tinge told us that daylight was in no hurry to leave our quiet valley. The mountain streams or creeks, which water so well the grassy plains among the Malvern Hills, are not affected to any considerable extent by dry summer weather. They are snow-fed from the high ranges, and each nor'-wester restores many a glacier or avalanche to its original form, and sends it flowing down the steep sides of yonder distant beautiful mountains to join the creeks, which, like a tangled skein of silver threads, ensure a good water supply to the New Zealand sheep-farmer. In the holes, under steep overhanging banks, the eels love to lurk, hiding from the sun's rays in cool depths, and coming out at night to feed. There are no fish whatever in the rivers, and I fear that the labours of the Acclimatization Society will be thrown away until they can persuade the streams themselves to remain in their beds like more civilised waters. At present not a month passes that one does not hear of some eccentric proceeding on the part of either rivers or creeks. Unless the fish are prepared to shift their liquid quarters at a moment's notice they will find themselves often left high and dry on the deserted shingle-bed. But eels are proverbially accustomed to adapt themselves to circumstances, and a fisherman may always count on getting some if he be patient.

About a mile down the flat, between very high banks, our principal creek ran, and to a quiet spot among the flax-bushes we directed our steps. By the fast-fading light the gentlemen set their lines in very primitive fashion. On the crumbling, rotten earth the New Zealand flax, the Phormium tenax, loves to grow, and to its long, ribbon-like leaves the eel-fishers fastened their lines securely, baiting each alternate hook with mutton and worms. I declared this was too cockney a method of fishing, and selected a tall slender flax-stick, the stalk of last year's spike of red honey-filled blossoms, and to this extempore rod I fastened my line and bait. When one considers that the old whalers were accustomed to use ropes made in the rudest fashion, from the fibre of this very plant, in their deep-sea fishing for very big prey, it is not surprising that we found it sufficiently strong for our purpose. I picked out, therefore, a comfortable spot,—that is to say, well in the centre of a young flax-bush, whose satiny leaves made the most elastic cushions around me; with my flax-stick held out over what was supposed to be a favourite haunt of the eels, and with Nettle asleep at my feet and a warm shawl close to my hand, prepared for my vigil. "Don't speak or move," were the gentlemen's last words: "the eels are all eyes and ears at this hour; they can almost hear you breathe." Each man then took up his position a few hundred yards away from me, so that I felt, to all intents and purposes, absolutely alone. I am "free to confess," as our American cousins say, that it was a very eerie sensation. It was now past ten o'clock; the darkness was intense, and the silence as deep as the darkness.

Hot as the day had been, the night air felt chill, and a heavy dew began to fall, showing me the wisdom of substituting woollen for cotton garments. I could see the dim outlines of the high hills, which shut in our happy valley on all sides, and the smell of the freshly-turned earth of a paddock near the house, which was in process of being broken up for English grass, came stealing towards me on the silent air. The melancholy cry of a bittern, or the shrill wail of the weka, startled me from time to time, but there was no other sound to break the eternal silence.

As I waited and watched, I thought, as every one must surely think, with strange paradoxical feelings, of one's own utter insignificance in creation, mingled with the delightful consciousness of our individual importance in the eyes of the Maker and Father of all. An atom among worlds, as one feels, sitting there at such an hour and in such a spot, still we remember with love and pride, that not a hair of our head falls to the ground unnoticed by an Infinite Love and an Eternal Providence. The soul tries to fly into the boundless regions of space and eternity, and to gaze upon other worlds, and other beings equally the object of the Great Creator's care; but her mortal wing soon droops and tires, and she is fain to nestle home again to her Saviour's arms, with the thought, "I am my Beloved's, and He is mine." That is the only safe beginning and end of all speculation. It was very solemn and beautiful, that long dark night,—a pause amid the bustle of every day cares and duties,—hours in which one takes counsel with one's own heart, and is still.

Midnight had come and gone, when the sputter and snap of striking a match, which sounded almost like a pistol shot amid the profound silence, told me that one of the sportsmen had been successful. I got up as softly as possible, wrapped my damp shawl round my still damper shoulders, and, fastening the flax-stick securely in the ground, stole along the bank of the creek towards the place where a blazing tussock, serving as a torch, showed the successful eel-fisher struggling with his prize. Through the gloom I saw another weird-looking figure running silently in the same direction; for the fact was, we were all so cramped and cold, and, weary of sitting waiting for bites which never came, that we hailed with delight a break in the monotony of our watch. It did not matter now how much noise we made (within moderate limits), for the peace of that portion of the creek was destroyed for the night. Half-a-dozen eels must have banded themselves together, and made a sudden and furious dash at the worsted ball, which Mr. U—— had been dangling in front of their mud hall-door for the last two hours. Just as he had intended, their long sharp teeth became entangled in the worsted loops, and although he declared some had broken away and escaped, three or four good-sized ones remained, struggling frantically.

It would have been almost impossible for one man to lift such a weight straight out of the water by a string; and as we came up and saw Mr. U— —'s agitated face in the fantastic flickering light of the blazing tussock, which he had set on fire as a signal of distress, I involuntarily thought of the old Joe Miller about the Tartar: "Why don't you let him go?" "Because he has caught me." It looked just like that. The furious splashing in the water below, and Mr. U— — grasping his line with desperate valour, but being gradually drawn nearer to the edge of the steep bank each instant. "Keep up a good light, but not too much," cried F— — to me, in a regular stage-whisper, as he rushed to the rescue. So I pulled up one tussock after another by its roots,—an exertion which resulted in upsetting me each time,—and lighted one as fast as its predecessor burned out. They were all rather damp, so they did not flare away too quickly. By the blaze of my grassy torches I saw F— —first seize Mr. U— — round the waist and drag him further from the bank; but the latter called out, "It's my hands,—they have no skin left: do catch hold, there's a good fellow." So the "good fellow" did catch hold, but he was too experienced an eel-fisher to try to lift a couple of dozen pounds weight of eels out of the water by a perpendicular string; so he tied it to a flax-bush near, and, stooping down in order to get some leverage over the bank, very soon drew the ball, with its slimy, wriggling captives, out of the water. Just as he jerked it far on shore, one or two of the creatures broke loose and escaped, leaving quite enough to afford a most disgusting and horrible sight as they were shuffled and poked into the empty flour-sack.

The sportsmen were delighted however, and departed to a fresh bend of the creek, leaving me to find my way back to my original post. This would have been difficult indeed, had not Nettle remained behind to guard my gloves, which I had left in his custody. As I passed, not knowing I was so near the spot, the little dog gave a low whimper of greeting, sufficient to attract my attention and guide me to where he was keeping his faithful watch and ward. I felt for my flax-stick and moved it ever so gently. A sudden jerk and splash startled me horribly, and warned me that I had disturbed an eel who was in the act of supping off my bait. In the momentary surprise I suppose I let go, for certain it is that the next instant my flax-stick was rapidly towed down the stream.

Instead of feeling provoked or mortified, it was the greatest relief to know that my eel-fishing was over for the night, and that now I had nothing to do except "wait till called for." So I took Nettle on my lap and tried to abide patiently, but I had not been long enough in New Zealand to have any confidence in the climate, and as I felt how damp my clothes were, and recollected with horror my West Indian experiences of the consequences of exposure to night air and heavy dew, my mind would dwell gloomily on the prospect of a fever, at least. It seemed a long and weary while before I perceived a figure coming towards me; and I am afraid I was both cross and cold and sleepy by the time we set our faces homewards. "I have only caught three," said F— —. "How many have you got?" "None, I am happy to say," I answered peevishly, "What could Nettle and I have done with the horrible things if we had caught any?"

The walk, or rather the stumble home, proved to be the worst part of the expedition. Not a ray of starlight had we to guide us,—nothing but inky blackness around and over us. We tried to make Nettle go first, intending to follow his lead, and trusting to his keeping the track; but Nettle's place was at my heels, and neither coaxing nor scolding would induce him to forego it. A forlorn hope was nothing to the dangers of each footstep. First one and then the other volunteered to lead the way, declaring they could find the track. All this time we were trying to strike the indistinct road among the tussocks, made by occasional wheels to our house, but the marks, never very distinct in daylight, became perfect will-o'-the-wisps at night. If we crossed a sheep-track we joyfully announced that we had found the way, but only to be undeceived the next moment by discovering that we were returning to the creek.

From time to time we fell into and over Spaniards, and what was left of our clothes and our flesh the wild Irishmen devoured. We must have got home somehow, or I should not be writing an account of it, at this moment, but really I hardly know how we reached the house. I recollect that the next day there was a great demand for gold-beater's skin, and court-plaster, and that whenever F—— and Mr. U—— had a spare moment during the ensuing week, they devoted themselves to performing surgical operations on each other with a needle; and that I felt very subdued and tired for a day or two. But there was no question of fever or cold, and I was stared at when I inquired whether it was not dangerous to be out all night in heavy dew after a broiling day.

We had the eels made into a pie by our shepherd, who assured me that if I entrusted them to my cook she would send me up such an oily dish that I should never be able to endure an eel again. He declared that the Maoris, who seem to have rather a horror of grease, had taught him how to cook both eels and wekas in such a way as to eliminate every particle of fat from both. I had no experience of the latter dish, but he certainly kept his word about the eels, for they were excellent.

Chapter III: Pig-stalking.

It was much too hot in summer to go after wild pigs. That was our winter's amusement, and very good sport it afforded us, besides the pleasure of knowing that we were really doing good service to the pastoral interest, by ridding the hills around us of almost the only enemies which the sheep have. If the squatter goes to look after his mob of ewes and lambs in the sheltered slopes at the back of his run, he is pretty nearly certain to find them attended by an old sow with a dozen babies at her heels. She will follow the sheep patiently from one camping ground to another, watching for a new-born and weakly lamb to linger behind the rest, and then she will seize and devour it. Besides this danger, the presence of pigs on the run keeps the sheep in an excited state. They have an uneasy consciousness that their foes are looking after them, and they move restlessly up and down the hills, not stopping to feed sufficiently to get fat. If a sheep-farmer thinks his sheep are not in good condition, one of the first questions he asks his shepherd is, "Are there any pigs about?" Our run had a good many of these troublesome visitors on it, especially in the winter, when they would travel down from the back country to grub up acres on acres of splendid sheep pasture in search of roots. The only good they do is to dig up the Spaniards for the sake of their delicious white fibres, and the fact of their being able to do this will give a better idea of the toughness of a wild pig's snout than anything else I can say.

It may be strange to English ears to hear a woman of tolerably peaceful disposition, and as the advertisements in the Times so often state, "thoroughly domesticated," aver that she found great pleasure in going after wild pigs; but the circumstances of the ease must be taken into consideration before I am condemned. First of all, it seemed terribly lonely at home if F—— was out with his rifle all day. Next, there was the temptation to spend those delicious hours of a New Zealand winter's day, between ten and four, out of doors, wandering over hills and exploring new gullies. And lastly, I had a firm idea that I was taking care of F——. And so I was in a certain sense, for if his rifle had burst, or any accident had happened to him, and he had been unable to reach the homestead, we should never have known where to find him, and days would probably have passed before every nook and corner of a run extending over many thousand acres could have been thoroughly searched.

I had heard terrible stories of shepherds slipping down and injuring themselves so that they could not move, and of their dead bodies being only found after weeks of careful seeking. F—— himself delighted to terrify me by descriptions of narrow escapes; and, as the pigs had to be killed, I resolved to follow in the hunter's train. The sport is conducted exactly like deer stalking, only it is much harder work, and a huge boar is not so picturesque an object as a stag of many tines, when you do catch sight of him. There is just the same accurate knowledge needed of the animal's habits and customs, and the same untiring patience. It is quite as necessary to be a good shot, for a grey pig standing under the lee of a boulder of exactly his own colour is a much more difficult object to hit from the opposite side of a ravine than a stag; and a wild boar is every whit as keen of scent and sharp of eye and ear as any antlered "Monarch of the Glen."

Imagine then a beautiful winter's morning without wind or rain. There has been perhaps a sharp frost over-night, but after a couple of hours of sunshine the air is as warm and bright as midsummer. We used to be glad enough of a wood fire at breakfast; but after that meal had been eaten we went into the verandah, open to the north-east (our warm quarter), which made a delicious winter parlour, and basked in the blazing sunshine. I used often to bring out a chair and a table, and work and read there all the morning, without either hat or jacket. But it sometimes happened that once or twice a week, on just such a lovely morning, F—— would proclaim his intention of going out to look for pigs, and, sooner than be left behind, I nearly always begged to be allowed to come too. There was no fear of my getting tired or lagging behind; and as I was willing to make myself generally useful, by carrying the telescope, a revolver for close quarters, and eke a few sandwiches, the offer of my company used to be graciously accepted. We could seldom procure the loan of a good pig-dog, and after one excursion with a certain dog of the name of "Pincher," I preferred going out by ourselves.

On that occasion F— — did not take his rifle, as there was no chance of getting a long shot at our game; for the dog would surely bring the pig to bay, and then the hunter must trust to a revolver or the colonial boar-spear, half a pair of shears (I suppose it should be called a shear) bound firmly on a flax stick by green flax-leaves. We had heard of pigs having been seen by our out-station shepherd at the back of the run, and as we were not encumbered by the heavy rifle, we mounted our horses and rode as far as we could towards the range where the pigs had been grubbing up the hill sides in unmolested security for some time past. Five miles from home the ground became so rough that our horses could go no further; we therefore jumped off, tied them to a flax-bush, taking off the saddles in case they broke loose, and proceeded on foot over the jungly, over-grown saddle. On the other side we came upon a beautiful gully, with a creek running through it, whose banks were so densely fringed with scrub that we could not get through to the stream, which we heard rippling amid the tangled shrubs. If we could only have reached the water our best plan would have been to get into it and follow its windings up the ravine; but even Pincher could hardly squeeze and burrow through the impenetrable fence of matapo and goi, which were woven together by fibres of a thorny creeper called "a lawyer" by the shepherds.

It was very tantalising, for in less than five minutes we heard trusty Pincher "speaking" to a boar, and knew that he had baled it up against a tree, and was calling to us to come and help him. F——ran about like a lunatic, calling out; "Coming Pincher: round him up, good dog!" and so forth; but they were all vain promises, for he could not get in. I did my best in searching for an opening, and gave many false hopes of having found one. At last I said, "If I run up the mountain side, and look down on that mass of scrub, perhaps I may see some way into it from above." "No: do you stay here, and see, if the pig breaks cover, which way he goes." Up the steep hill, therefore, F—— rushed, as swiftly and lightly as one of his own mountain sheep; and in a minute or two I saw him standing, revolver in hand, on an overhanging rock, peering anxiously down on the leafy mass below.

Pincher and the creek made such a noise between them that I could not hear what F—— said, and only guessed from his despairing gestures that there was no trap door visible in the green roof. I signalled as well as I could that he was to come down directly, for his-standing-place looked most insecure. Insecure indeed it proved. As I spoke the great fragment of rock loosely embedded in earth on the mountain side gave way with a crash, and came tumbling majestically down on the top of the scrub. As for F——, he described a series of somersaults in the air, which however agreeable in themselves, were very trying to the nerves of the spectatrix below. My first dread was least the rock should crush him, but to my great joy I saw at once that it was rolling slowly down the hill, whilst F——'s vigorous bound off it as it gave way, had carried him well into the middle of the leafy cushion beneath him, where he presently landed flat on his back!

I expected every moment to hear the revolver go off, but mercifully it did not do so; and as his thorny bed was hardly to be endured, F—— soon kicked himself off it, and before I could realize that he was unhurt, had scrambled to his feet, and was rushing off, crying in school-boy glee, "That will fetch him out" That (the rock) certainly did fetch him (the pig) out in a moment, and Pincher availed himself of the general confusion to seize hold of his enemy's hind leg, which he never afterwards let go. The boar kept snapping and champing his great tusks; but Pincher, even with the leg in his mouth, was too active to be caught: so as the boar found that it was both futile and undignified to try to run away with a dog hanging on his hind-quarters, he tried another plan. Making for a clump of Ti-ti palms he went to bay, and contrived to take up a very good defensive position. Pincher would have never given up his mouthful of leg if F—— had not called him off, for it seemed impossible to fire the revolver whilst the dog held on. This change of tactics was much against Pincher's judgment, and he kept rushing furiously in between F—— and the boar. As for me, I prudently retired behind a big boulder, on which I could climb if the worst came to the worst, and called out from time to time, to both dog and man, "Oh, don't!"

They did not even hear me, for the din of battle was loud. The pig dodged about so fast, that although F——'s bullets lodged in the palm tree at his back, not one struck a vulnerable part, and at last F——, casting his revolver behind him for me to pick up and reload, closed with his foe, armed only with the shear-spear. Pincher considered this too dangerous, and rushed in between them to distract the boar's attention. Just as F—— aimed a thrust at his chest,—for it was of no use trying to penetrate his hide,—the boar lowered his head, caught poor faithful Pincher's exposed flank, and tore it open with his razor-like tusk; but in the meantime the spear had gone well home into his brawny chest, exactly beneath the left shoulder, and his life-blood came gushing out. I was so infuriated at the sight of Pincher's frightful wound that I felt none of my usual pity for the victim; and rushing up to F—— with the revolver, of which only a couple of chambers were loaded, thrust it into his hand with an entreaty to "kill him quickly." This F—— was quite willing to do for his own sake, as a wounded boar is about the most dangerous beast on earth; and although the poor brute kept snapping at the broken flax-stick sticking in his heart, he fired a steady shot which brought the pig on his knees, only to roll over dead the next moment.

I cannot help pausing to say that I sewed up Pincher's wound then and there, with some of the contents of my Cambusmore house-wife; which always accompanied me on my sporting expeditions, and we carried him between us down to where the horses were fastened. There I mounted; and F— — lifting the faithful creature on my lap, we rode slowly home, dipping our handkerchiefs in cold water at every creek we crossed, and laying them on his poor flank. He was as patient and brave as possible, and bore his sufferings and weakness for days afterwards in a way which was a lesson to one, so grateful and gentle was he. His brave and sensible behaviour met its due reward in a complete though slow recovery.

I have only left myself space for one little sketch more; but it comes so vividly before me that I cannot shut it out. After a long day's walking, over the hills and vallies, so beautiful beneath our azure winter-sky, walking which was delightful as an expedition, but unsuccessful as to sport, we crossed the track of a large boar. We knew he was old by his being alone, and it was therefore very certain that he would show fight if we came up with him. Patiently we followed the track over a low saddle, through a clump of brushwood menuka, the broken twigs of which showed how large an animal had just passed by. Here and there a freshly grubbed-up Spaniard showed where he had paused for a snack; but at length we dropped down on the river bed, with its wide expanse of shingle, and there we lost all clue to our game.

After a little hesitation, F— — decided on climbing a high cliff on the right bank of the river, and trying to catch a glimpse of him. The opposite hill-side was gaunt and bare; a southern aspect shut out the sun in winter, and for all its rich traces of copper ore, "Holkam's Head" found no favour in the eyes of either shepherds or master. Grass would not grow there except in summer, and its gray, shingly sides were an eye-sore to its owner. We sat down on the cliff, and looked around carefully. Presently F— — said, in a breathless whisper of intense delight, "I see him." In vain I looked and looked, but nothing could my stupid eyes discover. "Lie down," said F— — to me, just as if I had been a dog. I crouched as low as possible, whilst F—

—settled himself comfortably flat on his stomach, and prepared to take a careful aim at the opposite side of the hill.

After what seemed a long time, he pulled his rifle's trigger, and the flash and crack was followed apparently by one of the gray boulders opposite leaping up, and then rolling heavily down the hill. F—— jumped up in triumph crying, "Come along, and don't forget the revolver." When we had crossed the river, reckless of getting wet to our waists in icy-cold water, F—— took the revolver from me and went first; but, after an instant's examination, he called out, "Dead as a door-nail! come and look at him." So I came, with great caution, and a more repulsive and disgusting sight cannot be imagined than the huge carcass of our victim already stiffening in death. The shot had been a fortunate one, for only an inch away from the hole the bullet had made his shoulders were regularly plated with thick horny scales, off which a revolver bullet would have glanced harmlessly, and he bore marks of having fought many and many a battle with younger rivals. His huge tusks were notched and broken, and he had evidently been driven out from among his fellows as a quarrelsome member of their society. Already the keen-eyed hawks were hovering above the great monster, and we left him to his fate in the solitary river gorge, where all was bleak and cold and gloomy,—a fitting death-place for the fierce old warrior.

Chapter IV: Skating in the back country.

I do not believe that even in Canada the skating can be better than that which was within our reach in the Malvern Hills. Among our sheltered valleys an sunny slopes the hardest frost only lasted a few hour after dawn; but twenty-five miles further back, on the border of the glacier region, the mountain tarns could boast of ice several feet thick all the winter. We heard rumours of far-inland lakes, across which heavily-laden bullock-teams could pass in perfect safety for three months of the year, and we grumbled at the light film over our own large ponds, which would not bear even my little terrier's weight after mid-day: and yet it was cold enough at night, during our short bright winters, to satisfy the most icy-minded person. I think I have mentioned before that the wooden houses in New Zealand, especially those roughly put together up-country, are by no means weather-tight. Disagreeable as this may be, it is doubtless the reason of the extraordinary immunity from colds and coughs which we hill-dwellers enjoyed. Living between walls formed by inch-boards over-lapping each other, and which can only be made to resemble English rooms by being canvassed and papered inside, the pure fresh air finds its way in on all sides. A hot room in winter is an impossibility, in spite of drawn curtains and blazing fires, therefore the risk of sudden changes of temperature is avoided.

Some such theory as this is absolutely necessary to account for the wonderfully good health enjoyed by all, in the most capricious and trying climate I have ever come across. When a strong nor'-wester was howling down the glen, I have seen the pictures on my drawing-room walls blowing out to an angle of 45 degrees, although every door and window in the little low wooden structure had been carefully closed for hours. It has happened to me more than once, on getting up in the morning, to find my clothes, which had been laid on a chair beneath my bedroom window overnight, completely covered by powdered snow, drifting in through the ill-fitting casement. This same window was within a couple of feet of my bed, and between me and it was neither curtain nor shelter of any sort. Of a winter's evening I have often been obliged to wrap myself up in a big

Scotch maud, as I sat, dressed in a high linsey gown, by a blazing fire, so hard was the frost outside; but by ten o'clock next morning I would be loitering about the verandah, basking in the sunshine, and watching the light flecks of cloud-wreaths and veils floating against an Italian-blue sky. Yet such is the inherent discontent of the human heart, that instead of rejoicing in this lovely mid-day sunshine, we actually mourned over the vanished ice which at daylight had been found, by a much-envied early riser, strong enough to slide on for half an hour. It seemed almost impossible to believe that any one had been sliding that morning within a few feet of where I sat working in a blaze of sunshine, with my pretty grey and pink Australian parrot pluming itself on the branch of a silver wattle close by, and "Joey," the tiny monkey from Panama, sitting on the skirt of my gown, with a piece of its folds arranged by himself shawl-wise over his glossy black shoulders. If either of these tropical pets had been left out after four o'clock that sunny day, they, would have been frozen to death before our supper time.

It was just on such a day as this, and in just such a bright mid-day hour, that a distant neighbour of ours rode up to the garden gate, leading a pack horse. Outside the saddle-bags, with which this animal was somewhat heavily laden, could be plainly seen a beautiful new pair of Oxford skates, glinting in the sunshine; and it must have been the sight of these beloved implements which called forth the half-envious remark from one of the gentlemen, "I suppose you have lots of skating up at your place?"

"Well, not exactly at my station, but there is a capital lake ten miles from my house where I am sure of a good day's skating any time between June and August," answered Mr. C. H— —, our newly arrived guest.

We all looked at each other. I believe I heaved a deep sigh, and dropped my thimble, which "Joey" instantly seized, and with a low chirrup of intense delight, commenced to poke down between the boards of the verandah. It was too bad of us to give such broad hints by looks if not by words. Poor Mr. C. H— — was a bachelor in those days: he had not been at his little out-of-the-way homestead for some weeks, and was ignorant of its resources in the way of firing (always an important matter at a station), or even of tea and mutton. He had no woman-servant, and was totally unprepared for an incursion of skaters; and yet,—New Zealand fashion,—no sooner did he perceive that we were all longing and pining for some skating, than he invited us all most cordially to go up to his back-country run the very next day, with him, and skate as long as we liked. This was indeed a delightful prospect, the more especially as it happened to be only Monday, which gave us plenty of time to be back again by Sunday, for our weekly service. We made it a rule never to be away from home on that day, lest any of our distant congregation should ride their twenty miles or so across country and find us absent.

When the host is willing and the guests eager, it does not take long to arrange a plan, so the next morning found three of us, besides Mr. C. H—— mounted and ready to start directly after breakfast. I have often been asked how I managed in those days about toilette arrangements, when it was impossible to carry any luggage except a small "swag," closely packed in a waterproof case and fastened on the same side as the saddle-pocket. First of all I must assure my lady readers that I prided myself on turning out as neat and natty as possible at the end of the journey, and yet I rode not only in my every-day linsey gown, which could be made long or short at pleasure, but in my crinoline. This was artfully looped up on the right side and tied by a ribbon, in such a way that when I came out ready dressed to mount, no one in the world could have guessed that I had on any cage beneath my short riding habit with a loose tweed jacket over the body of the dress. Within the "swag" was stowed a brush and comb, collar, cuffs and handkerchiefs, a little necessary linen, a pair of shoes, and perhaps a ribbon for my hair if I meant to be very smart. On this occasion we all found that our skates occupied a terribly large proportion both of weight and space in our modest kits, but still we were much too happy to grumble.

Where could you find a gayer quartette than started at an easy canter up the valley that fresh bracing morning? From the very first our faces were turned to the south-west, and before us rose the magnificent chain of the Southern Alps, with their bold snowy peaks standing out in a glorious dazzle against the cobalt sky. A stranger, or colonially speaking, a "new chum," would have thought we must needs cross that barrier-range before we could penetrate any distance into the back country, but we knew of long winding vallies and gullies running up between the giant slopes, which would lead us, almost without our knowing how high we had climbed, up to the elevated but sheltered plateau among the back country ranges where Mr. C. H— —'s homestead stood. There was only one steep saddle to be crossed, and that lay between us and Rockwood, six miles off. It was the worst part of the journey for the horses, so we had easy consciences in dismounting and waiting an hour when we reached that most charming and hospitable of houses. I had just time for one turn round the beautiful garden, where the flowers and shrubs of old England grew side by side with the wild and lovely blossoms of our new island home, when the expected coo-e rang out shrill and clear from the rose-covered porch. It was but little past mid-day when we made our second start, and set seriously to work over fifteen miles of fairly good galloping ground. This distance brought us well up to the foot of a high range, and the last six miles of the journey had to be accomplished in single file, and with great care and discretion, for the track led through bleak desolate vallies, round the shoulder of abutting spurs, through swamps, and up and down rocky staircases. Mr. C. H— — and his cob both knew the way well however, and my bay mare Helen had the cleverest legs and the wisest as well as prettiest head of her race. If left to herself she seldom made a mistake, and the few tumbles she and I ever had together, took place only when she found herself obliged to go my way instead of her own. We entered the gorges of the high mountains between us and the west, and soon lost the sun; even the brief winter twilight faded away more swiftly than usual amid those dark defiles; and it was pitch dark, though only five o'clock, when we heard a sudden and welcome clamour of dog voices.

These deep-mouthed tones invariably constitute the first notes of a sheep-station's welcome; and a delightful sound it is to the belated and bewildered traveller, for besides guiding his horse to the right spot, the noise serves to bring out some one to see who the traveller may be. On this occasion we heard one man say to the other, "It's the boss:" so almost before we had time to dismount from our tired horses (remember they had each carried a heavy "swag" besides their riders), lights gleamed from the windows of the little house, and a wood fire sparkled and sputtered on the open hearth. Mr. C. H— — only just guided me to the door of the sitting-room, making an apology and injunction together,—"Its very rough I am afraid: but you can do what you like;"—before he hastened back to assist his guests in settling their horses comfortably for the night. Labour used to be so dear and wages so high, especially in the back country of New Zealand, that the couple of men,—one for indoor work, to saw wood, milk, cook, sweep, wash, etc., and the other to act as gardener, groom, ploughman, and do all the numerous odd jobs about a place a hundred miles and more from the nearest shop,—represented a wage-expenditure of at least 200 pounds a year. Every gentleman therefore as a matter of course sees to his own horse when he arrives unexpectedly at a station, and I knew I should have at least half an hour to myself.

The first thing to do was to let down my crinoline, for I could only walk like a crab in it when it was fastened up for riding, kilt up my linsey gown, take off my hat and jacket, and set to work The curtains must be drawn close, and the chairs moved out from their symmetrical positions against the wall; then I made an expedition into the kitchen, and won the heart of the stalwart cook, who was already frying chops over the fire, by saying in my best German, "I have come to help you with the tea." Poor man! it was very unfair, for Mr. C. H— — had told me during our ride that his servitor was a German, and I had employed the last long hour of the journey in rubbing up my exceedingly rusty knowledge of that language, and arranging one or two effective sentences. Poor Karl's surprise and delight knew no bounds, and he burst forth into a long monologue, to which I could find no readier answers than smiles and nods, hiding my inability to follow up my brilliant beginning under the pretence of being very busy. By the time the gentlemen had stabled and fed the horses and were ready, Karl and I between us had arranged a bright cosy little apartment with a capital tea-dinner on the table. After this meal there were pipes and toddy, and as I could not retire, like Mrs. Micawber at David Copperfield's supper party, into the adjoining bedroom and sit by myself in the cold, I made the best of the somewhat dense clouds of smoke with which I was soon surrounded, and listened to the fragmentary plans for the next day. Then we all separated for the night, and in two minutes I was fast asleep in a little room no bigger than the cabin of a ship, with an opossum rug on a sofa for my bed and bedding.

It was cold enough the next morning, I assure you: so cold that it was difficult to believe the statement that all the gentlemen had been down at daybreak to bathe in the great lake which spread like an inland sea before the bay-window of the little sitting room. This lake, the largest of the mountain chain, never freezes, on account partly of its great depth, and also because of its sunny aspect. Our destination lay far inland, and if we meant to have a good long day's skating we must start at once. Such a perfect day as it was! I felt half inclined to beg off the first day on the ice, and to spend my morning wandering along the rata-fringed shores of Lake Coleridge, with its glorious enclosing of hills which might fairly be called mountains; but I feared to seem capricious or lazy, when really my only difficulty was in selecting a pleasure. The sun had climbed well over the high barriers which lay eastwards, and was shining brightly down through the quivering blue ether overhead; the frost sparkled on every broad flax-blade or slender tussock-spine, as if the silver side of earth were turned outwards that winter morning.

No sooner had we mounted (with no "swag" except our skates this time) than Mr. C. H— — set spurs to his horse, and bounded over the slip-rail of the paddock before Karl could get it down. We were too primitive for gates in those parts: they only belonged to the civilization nearer Christchurch; and I had much ado to prevent my pony from following his lead, especially as the other gentlemen were only too delighted to get rid of some of their high spirits by a jump. However Karl got the top rail down for me, and "Mouse" hopped over the lower one gaily, overtaking the leader of the expedition in a very few strides. We could not keep up our rapid pace long; for the ground became terribly broken and cut up by swamps, quicksands, blind creeks, and all sorts of snares and pit-falls. Every moment added to the desolate grandeur of the scene. Bleak hills rose up on either hand, with still bleaker and higher peaks appearing beyond them again. An awful silence, unbroken by the familiar cheerful sound of the sheep calling to each other,—for even the hardy merino cannot live in these ranges during the winter months,—brooded around us, and the dark mass of a splendid "bush," extending over many hundred acres, only added to the lonely grandeur of the scene. We rode almost the whole time in a deep cold shade, for between us and the warm sun-rays were such lofty mountains that it was only for a few brief noontide moments he could peep over their steep sides.

After two hour's riding, at the best pace which we could keep up through these terrible gorges, a sharp turn of the track brought us full in view of our destination. I can never forget that first glimpse of Lake Ida. In the cleft of a huge, gaunt, bare hill, divided as if by a giant hand, lay a large black sheet of ice. No ray of sunshine ever struck it from autumn until spring, and it seemed impossible to imagine our venturing to skate merrily in such a sombre looking spot. But New-Zealand sheep farmers are not sentimental I am afraid. Beyond a rapid thought of self-congratulation that such "cold country" was not on their run, they did not feel affected by its eternal silence and gloom. The ice would bear, and what more could skater's heart desire? At the end of the dark tarn, nearest to the track by which we had approached it, stood a neat little hut; and judge of my amazement when, as we rode up to it, a young gentleman, looking as if he was just going out for a day's deer-stalking, opened the low door and came out to greet us. Yes, here was one of those strange anomalies peculiar to the colonies. A young man, fresh from his University, of refined tastes and cultivated intellect, was leading here the life of a boor, without companionship or appreciation of any sort. His "mate" seemed to be a rough West countryman, honest and well meaning enough, but utterly unsuited to Mr. K——. It was the old story, of wild unpractical ideas hastily carried out. Mr. K—— had arrived in New Zealand a couple of years before, with all his worldly wealth,—1,000 pounds. Finding this would not go very far in the purchase of a good sheep-run, and hearing some calculations about the profit to be derived from breeding cattle, based upon somebody's lucky speculation, he eagerly caught at one of the many offers showered upon unfortunate "new chums," and bought the worst and bleakest bit of one of the worst and bleakest runs in the province. The remainder of his money was laid out in purchasing stock; and now he had sat down patiently to await, in his little hut, until such time as his brilliant expectations would be realized. I may say here they became fainter and fainter year by year, and at last faded away altogether; leaving him at the end of three lonely, dreadful years with exactly half his capital, but double his experience. However this has nothing to do with my story, except that I can never think of our skating expedition to that lonely lake, far back among those terrible hills, without a thrill of compassion for the only living human being, who dwelt among them.

It was too cold to dawdle about, however, that day. The frost lay white and hard upon the ground, and we felt that we were cruel in leaving our poor horses standing to get chilled whilst we amused ourselves. Although my beloved Helen was not there, having been exchanged for the day in favour of Master Mouse, a shaggy pony, whose paces were as rough as its coat, I begged a red blanket from Mr. K— —, and covered up Helen's stable companion, whose sleek skin spoke of a milder temperature than that on Lake Ida's "gloomy shore." Our simple arrangements were soon made. Mr. K— — left directions to his mate to prepare a repast consisting of tea, bread, and mutton for us, and, each carrying our skates, we made the best of our way across the frozen tussocks to the lake. Mr. K— — proved an admirable guide over its surface, for he was in the habit during the winter of getting all his firewood out of the opposite "bush," and bringing it across the lake on sledges drawn by bullocks. We accused him of having cut up our ice dreadfully by these means; but he took us to a part of the vast expanse where an unbroken field of at least ten acres of ice stretched smoothly before us. Here were no boards marked "DANGEROUS," nor any intimation of the depth of water beneath. The most timid person could feel no apprehension on ice which seemed more solid than the earth; so accordingly in a few moments we had buckled and strapped on our skates, and were skimming and gliding—and I must add, falling—in all directions. We were very much out of practice at first, except Mr. K— —, who skated every day, taking short cuts across the lake to track a stray heifer or explore a blind gully.

I despair of making my readers see the scene as I saw it, or of conveying any adequate idea of the intense, the appalling loneliness of the spot. It really seemed to me as if our voices and laughter, so far from breaking the deep eternal silence, only brought it out into stronger relief. On either hand rose up, shear from the waters edge, a great, barren, shingly mountain; before us loomed a dark pine forest, whose black shadows crept up until they merged in the deep crevasses and fissures of the Snowy Range. Behind us stretched the winding gullies by which we had climbed to this mountain tarn, and Mr. K——'s little hut and scrap of a garden and paddock gave the one touch of life, or possibility of life, to this desolate region. In spite of all scenic wet blankets we tried hard to be gay, and no one but myself would acknowledge that we found the lonely grandeur of our "rink" too much for us. We skated away perseveringly until we were both tired and hungry, when we returned to Mr. K——'s hut, took a hasty meal, and mounted our chilled steeds. Mr. C. H—— insisted on bringing poor Mr. K—— back with us, though he was somewhat reluctant to come, alleging that a few days spent in the society of his kind made the solitude of his weather-board hut all the more dreary. The next day and yet the next we returned to our gloomy skating ground, and when I turned round in my saddle as we rode away on Friday evening, for a last look at Lake Ida lying behind us in her winter black numbness, her aspect seemed more forbidding than ever, for only the bare steep hill-sides could be seen; the pine forest and white distant mountains were all blotted and blurred out of sight by a heavy pall of cloud creeping slowly up.

"Let us ride fast," cried Mr. K——, "or we shall have a sou'-wester upon us;" so we galloped home as quickly as we could, over ground that I don't really believe I could summon courage to walk across, ever so slowly, to-day,—but then one's nerves and courage are in very different order out in New Zealand to the low standard which rules for ladies in England, who "live at home in ease!" Long before we reached home the storm was pelting us: my little jacket was like a white board when I took it off, for the sleet and snow had frozen as it fell. I was wet to the skin, and so numb with cold I could hardly stand when we reached home at last in the dark and downpour. I could only get my things very imperfectly dried, and had to manage as best I could, but yet no one even thought of making the inquiry next morning when I came out to breakfast, "Have you caught cold?" It would have seemed a ridiculous question.

Chapter V: Toboggon-ing.

I cannot resist the temptation to touch upon one of the winter amusements which came to us two years later. Yet the word "amusement" seems out of place, no one in the Province having much heart to amuse themselves, for the great snow storm of August, 1867, had just taken place, and we were in the first days of bewilderment at the calamity which had befallen us all. A week's incessant snow-fall, accompanied by a fierce and freezing south-west wind, had not only covered the whole of the mountains from base to brow with shining white, through which not a single dark rock jutted, but had drifted on the plains for many feet deep. Gullies had been filled up by the soft, driving flakes, creeks were bridged over, and for three weeks and more all communication between the stations and the various townships was cut off. The full extent of our losses was unknown to us, and dreary as were our forebodings of misfortune, none of us guessed that snow to be the winding sheet of half a million of sheep. The magnificent semi-circle of the Southern Alps stood out, for a hundred miles from north to south, in appalling white distinctness, and no one in the whole Colony had ever seen the splendid range thus free from fleck or flaw. We had done all we could within working distance, but what was, the use of digging in drifts thirty feet deep? Amidst, and almost above, the terrible anxiety about our own individual safety,—for the snow was over the roof of many of the station-houses,—came the pressing question, "Where are the sheep?" A profound silence unbroken by bleat of lamb, or bark of dog, or any sound of life, had reigned for many days, when a merciful north-westerly gale sprung, up, and releasing the heavily-laden earth from its white bondage, freed the miserable remnant of our flocks and herds. At least, I should say, it freed those sheep which had travelled down to the vallies, driven before the first pitiless gusts, but we knew that many hundreds, if not thousands, of wethers must have been surprised and imprisoned far back among the hills.

Such knowledge could not be acted upon, however, for no human being could hope to plunge through the drifts around us. Old shepherds who had lived on the run for fifteen years, confessed that they did not know their way fifty yards from the homestead. The vallies were filled up, so that one gully looked precisely like its fellow; rocks, scrub, Ti-ti palms, all our local land-marks had disappeared; not a fence or gate could be seen in all the country side. Here and there a long wave-like line in the smooth mass would lead us to suppose that a wire fence lay buried beneath its curves, but we had no means of knowing for certain. Near the house every shrub and out-building, every hay-stack or wood-heap, had all been covered up, and no man might even guess where they lay.

This had been the terrible state of things, and although the blessed warm wind had removed our immediate and pressing fear of starvation, we could not hope to employ ourselves in searching for our missing sheep for many days to come. None of us had been able to take any exercise for more than a fortnight, and having done all that could possibly be done near at hand, F— — set to work to manufacture some sledges out of old packing-cases. Quite close to the house, a hill sloped smoothly for about 300 yards, at an angle of 40 degrees; along its side lay a perfectly level and deep drift, which did not show any signs of thawing for more than a month, and we resolved to use this as a natural Montagne Russe. The construction of a suitable sledge was the first difficulty to be surmounted, and many were the dismal failures and break-neck catastrophes which preceded what we considered a safe and successful vehicle. Not only was it immensely difficult to make, without either proper materials or tools, a sledge which could hold two people (for F— — declared it was no fun sleighing alone), but his "patent brakes" proved the most broken of reeds to lean upon when the sledge was dashing down the steep incline at the rate of a thousand miles an hour.

We nearly broke our necks more than once, and I look back now with amazement to our fool-hardiness. How well I remember one expedition, when F— —, who had been hammering away in a shed all the morning, came to find me sitting in the sun in the verandah, and to inform me that at last he had perfected a conveyance which would combine speed with safety. Undaunted by previous mishaps, I sallied forth, and in company with Mr. U— — and F— —, climbed painfully up the high hill I have mentioned, by some steps which they had cut in the frozen snow. Without some such help we could not have kept our footing for a moment, and as long as I live I shall never forget the sensation of leaving my friendly Alpenstock planted in the snow, and of seating myself on that frail sledge. Perhaps I ought to describe it here. A board, about six feet long by one foot broad, with sheet-iron nailed beneath it, and curved upwards in front; on its upper surface a couple of battens were fixed, one quite at the foremost end, and one half-way. That was F— —'s new patent sledge, warranted to go faster down an incline than any other conveyance on the surface of the earth. I was the wretched "passenger," as he called me, on more than one occasion, and I will briefly describe my experiences. "Why did you go?" is a very natural question to arise in my reader's mind; and sitting here at my writing-table, I feel as if I must have been a lunatic to venture. But in those delicious wild days, no enterprise seemed too rash or dangerous to engage in, from mounting a horse which had never seen or felt the fluttering of a habit, to embarking on the conveyance I have described above, and starting down a mountain-side at the risk of a broken neck.

Well, to return to that terrible moment. I see the whole scene now. The frail, rude sledge, with its breaks made out of a couple of standards from a wire fence, connected by a strong iron chain; F— —seated at the back of the precious contrivance, firmly grasping a standard in each hand; Mr. U— — clinging desperately to his Alpen-stock with one hand, whilst with the other he helps me on to the board; and Nettle, my dear little terrier, standing shivering on three legs, sniffing distrustfully at the sledge. It is extremely difficult even to take one's place on a board a dozen inches wide. My petticoats have to be firmly wrapped around me, and care taken that no fold projects beyond the sledge, or I should be soon dragged out of my frail seat. I fix my feet firmly against the batten, and F— — cries, "Are you ready?" "Oh, not yet!" I gasp, clinging to Mr. U— —'s hand as if I never meant to let it go. "Hold tight!" he shouts. Now what a mockery this injunction was. I had nothing to hold on to except my own knees, and I clasped them convulsively. Mr. U— — says, "You're all right now," and before I can realize that he has let go my hand, before my courage is half-way up to the necessary height, we are off. The breaks are slightly depressed for the first few yards, in order to regulate our pace, and because there is a tremendously steep pitch just at first. Once we have safely passed that he tilts up the standards, and our sledge shoots like a meteor down the perfectly smooth incline. I cannot draw my breath, we are going at such a pace through the keen air; I give myself up for lost. We come to another steep pitch near the bottom of the hill; F— — is laughing to such a degree at me that he does not put down his breaks soon enough, and loses control of the sledge. We appear to leap down the dip, and then the sledge turns first one way and then the other, its zinc prow being sometimes up-hill and some-times down. It seems wonderful that we keep on the sledge, for we have no means of holding on except by pressing our feet against the battens; yet in the grand and final upset at the bottom of the hill, the sledge is there too, and we find we have never parted company from it.

Will any one believe that after such a perilous journey, I could actually be persuaded to try again? But so it was. At first the fright (for I was really terrified) used to make me very cross, and I declared that I was severely hurt, if not "kilt entirely;" but after I had shaken the snow out of my linsey skirt, and discovered that beyond the damage to my nerves I was uninjured, F— — was quite sure to try to persuade me to make another attempt, and I was equally sure to yield to the temptation. As well as my memory serves me, we only made one really successful journey, and that was on an occasion when we kept the breaks down the whole way. But I never could insure similar precautions being taken again, and we consequently experienced every variety of mishaps possible to sledge travellers. I persevered however for some days until the north-westerly wind, which was blowing softly all the time, began to lay bare the sharpest points of the rocks, and then I gave in at once, and would not be a "passenger" any more. It was rather too much to strike one's head against a jagged fragment of rock, or to dislocate one's thumb against a concealed stump of a palm tree. Then the sharp points of the Spaniards began to stick up through the softening snow, and nothing would induce me to run the risk of touching their green bayonets. Besides which, the fast-thawing snow made it very difficult to climb up to the top of our hill, for the carefully-cut steps had disappeared long ago. So I gave up sledge journeys on my own account, and used only to look at F— — and Mr. U— — taking them.

These two persevered so long as an inch of snow remained on the hill-side. Some of their adventures were very alarming, and certainly rather dangerous. One afternoon I had been watching them for more than an hour, and had seen them go through every variety of disaster, and capsize with no further effect than increasing their desire for "one more" trial. On the blind-side of the hill,—that is to say the side which gets scarcely any sun in winter,—a deep drift of snow still lingered, filling up a furrow made in former years by a shingle-slip. Thither the two adventurous climbers dragged their sledge, and down the steep incline they performed their perilous descent many a time. I became tired of watching the board shoot swiftly over the white streak; and I strolled round the shoulder of the hill, to see if there was any appearance of the snow-fall lessening in the back country.

I must have been away about half an hour, and had made the circuit of the little knoll which projected from the mountain side, returning to where I expected to find sleigh and sleighers starting perhaps on just "one more" journey. But no one was there, and a dozen yards or so from the usual starting-point, the snow was a good deal ploughed up and stained in large patches by blood. Here was an alarming spectacle, though the only wonder was that a bad accident had not occurred before. I saw the sledge, deserted and broken, near the end of the drift: of the passengers there was neither sign nor token. I must say I was terribly frightened, but it is useless in New Zealand to scream or faint; the only thing to do in an emergency is to coo-e; and so, although my heart was thumping loudly in my ears, and at first I could not produce a sound, I managed at last, after many attempts, to muster up a loud clear coo-e. There was the usual pause, whilst the last sharp note rang back from the hill-sides, and vibrated through the clear silent air; and then, oh, welcome sound! I heard a vigorous answer from our own flat where the homestead stood. I set off down-hill as fast as I could, and had the joy, when I turned the slope which had hidden our little house from my view, to see F— — and Mr. U— — walking about; but even from that distance I could see that poor Mr. U— —'s head was bandaged up, and as soon as I got near enough to hear, F— —shouted "I have broken my neck!" adding, "I am very hungry: let us go in to supper."

Under the circumstances these words were consolatory; and when I came to hear the story, this was the way the accident happened. As I mentioned before, even this drift had thawed till it was soft at the surface and worn away almost to the rocks. During a rapid descent the nose of the sledge dipped through the snow, and stopped dead against a rock. Mr. U— — was instantly buried in the snow, falling into a young but prickly Spaniard, which assaulted him grievously; but F— — shot over his head some ten yards, turned a somersault, and alit on his feet. This sounds a harmless performance enough, but it requires practice; and F— — declared that for weeks afterwards his neck felt twisted. The accident must have looked very ridiculous: the sledge one moment gliding smoothly along at the rate of forty miles an hour,—the next a dead stop, and F— —flying through the air over his passenger's head, finishing feet first plump down in the soft snow.

Looking back on that time, I can remember how curiously soon the external traces of the great snow-storm disappeared. For some weeks after the friendly nor-wester, the air of the whole neighbourhood was tainted by dead and decaying sheep and lambs; and the wire fences, stock-yard rails, and every "coign of vantage," had to be made useful but ghastly by a tapestry of sheep-skins. The only wonder was that a single sheep had survived a storm severe enough to kill wild pigs. Great boars, cased in hides an inch thick, had perished through sheer stress of weather; while thin-skinned animals, with only a few months growth of fine merino wool on their backs, had endured it all. It was well known that the actual destruction of sheep was mainly owing to the two days of heavy rain which succeeded the snow. Out of a flock of 13,000 of all ages, we lost, on the lowest calculation, 1,000 grown sheep and nearly 3,000 lambs; and yet our loss was small by comparison with that of our neighbours, whose runs were further back among the hill, and less sheltered than our own.

Long before midsummer our cloud-shadowed hills were green once more; and I think I see again their beautiful outlines, their steep sides planted with semi-tropical palms and grasses, whilst the more distant peaks are veiled in a sultry haze. During that peculiarly bright and lovely summer we often ask each other, Could it have been true that no one knew one mountain from the other, and that hills had been apparently levelled and vallies filled up by the heaviest snow-fall ever known. But whilst the words were on our lips, we could see a group of palm-trees, ten feet high, with their topmost leaves gnawed to the stump by starving sheep, that must have been standing on at least seven feet of snow to reach them; and there was scarcely a creek on the run whose banks were not strewn, for many a long day, by bare and bleaching bones.

Chapter VI: Buying a run.

Like many other people in the world, I have occasionally built castles in the air, and equally of course they have invariably tumbled down in due time with a crash This particular castle however, not only attained to a great elevation in the visionary builder's eyes, but it covered so vast an area of land, that the story of its rise and fall deserves to be placed on record, as a warning to aerial architects and also as a beacon-light to young colonists.

This was exactly the way it all happened. The new year of 186-found us living very quietly and happily on a small compact sheep-farm, at the foot of the Malvern Hills, in the province of Canterbury, New Zealand. As runs went, its dimensions were small indeed; for we only measured it at 12,000 acres, all told. The great tidal wave of prosperity, which sets once in a while towards the shores of all colonies, had that year swelled and risen to its full force; but this we did not know. Borne aloft upon its unsubstantial crest we could not, from that giddy height, discern any water-valleys of adversity or clouds of change and storm along the shining horizon of the new world around us. All our calculations were based on the assumption that the existing prices for sheep, wool, cattle, and all farm-produce, would rule for many a long day; and the delightful part of this royal road to wealth was, that its travellers need not exert themselves in any way: they had only to sit still with folded hands whilst their sheep increased, and it was well known that a flock doubled itself in three short years. The obvious deduction from this agreeable numerical fact was, that in an equally short period your agent's payments to your bank account would also be doubled. In the meantime the drays were busy carting the wool to the seaports as fast as they could be loaded, whilst speculative drovers rode all about the country buying up the fat cattle and wethers from every run. These were wanted to supply the West Coast Diggings which had just "broken out" (as the curious phrase goes there), and so was every description of grain and dairy produce.

We squatters were not the only inhabitants of this fool's paradise. The local Government began planning extensive works: railways were laid out in every direction, bridges planned across rivers, which proved the despair of engineers; whilst a tunnel, the wonder of the Southern Hemisphere, was commenced through a range of hills lying between Port Lyttleton and Christchurch. All this work was undertaken on a scale of pay which made the poor immigrants who thronged to the place by every ship, rub their eyes and believe they must be dreaming, and that they would presently wake up and find themselves back again in the old country, at the old starvation rate of wages. Small capitalists, with perhaps only one or two hundred pounds in the world, bid against each other as purchasers of quarter-acre sections in the fast-springing townships, or of fifty-acre lots of arable land in the projected suburbs. Subscriptions were raised for building a Cathedral in Christchurch; but so dear was both labour and material, that 7,000 pounds barely sufficed to lay its foundations.

The paramount anxiety in men's minds seemed to be to secure land. Sheep-runs in sheltered accessible parts of the country commanded enormous prices, and were bought in the most complicated way. The first comers had taken up vast tracts of land in all directions from the Government, at an almost nominal rental. This had happened quite in the dark and remote ages of the history of the colony, at least ten or twelve years before the date of which I write. As speculators with plenty of hard cash came down from Australia, these original tenants sold, as it were, the good-will and stock of their run at enormous prices; but what always seemed to me so hard was, that after you had paid any number of thousand pounds for your run, you might have to buy it all, or at any rate, some portion of it, over again. Land could only be purchased freehold from the Government, for 2 pounds an acre; and if a "cockatoo" (i.e., a small farmer), or a speculator in mines, fancied any part of your property, he had only to go to the land office, and challenge your pre-emptive rights. The officials gave you notice of the challenge, and six weeks' grace in which to raise the money, and buy it freehold yourself; but few sheep-farmers could afford to pay a good many hundred pounds unexpectedly to secure even their best "flats" or vallies. Hence it often happened that large runs in the most favourable situations were cut up by small investors, "free selectors" as they are called in Australia, and it used to be rather absurd the way one grew to distrust any stranger who was descried riding about the run. The poor man might be looking for a stray horse, or have lost his way, but we always fancied he must be "prospecting" for either gold or coals, or else be a "cockatoo" disguised as a traveller.

Such was the state of things when my story opens. Shearing was just over, and we knew to a lamb how rapidly our flocks and herds were increasing. A succession of mild winters and early genial springs had got the flock into capital order. The wool had all been sent off to Christchurch by drays, the sheep were turned out on the beautiful green hills for ten months of perfect rest and peace; whilst the dogs, who had barked themselves quite hoarse, were enabled to desist from their labours in mustering and watching the yet unshorn mobs on the vallies. Although our run was as well grassed and watered as any in the province, still it could not possibly carry more than a certain number of sheep, and to that total our returns showed that we were rapidly approaching. The most careful calculations warned us that by next shearing we should hardly know what to do with our sheep. It is always better to be under than overstocked, for the merino gets out of condition immediately, and even the staple of the wool deteriorates if its wearer be at all crowded on his feeding-grounds.

"You must take up more country directly," was the invariable formula of the advice we, comparatively "new chums," received on all sides. This was easier to say than to do. Turn which ever way we would, far back beyond our own lovely vallies and green hills, back up to the bleak region of glaciers, where miles of bush and hundreds of acres of steep hill-side, formed the back-est of "back country," every inch of land was taken up. No fear had those distant Squatters of "cockatoos," or even of miners; for no one came their way who could possibly help it. Still we should have been comparatively glad to buy such a run fifty or sixty miles further back,—at the foot, in fact of the great Southern Alps,—just as a summer feeding-ground for the least valuable portion of our flock. But no one was inclined to part with a single acre, and we were forced to turn our eyes in a totally different direction.

If my readers will refer to the accompanying map of New Zealand, and look at the Middle or South Island, they will notice a long seaboard on the eastern side of the island, stretching SS.W. for many hundred leagues. It extends beyond the Province of Canterbury to that of Otago, and embraces some of the most magnificent pastoral land in the settlement. Not only is the soil rich and productive, but the climate is rather less windy than with us in the northern portion of the island; and the capital of Otago (Dunedin) had risen into comparative position and importance before Christchurch,— was in short an elder sister of that pretty little town. Most of the settlers in Otago were Scotchmen, and as there are no better colonists anywhere, its prosperity had attained to a very flourishing height. Gold-digging had also broken out at the foot of the Dunstan range, so that Otago held her head quite as high, if not higher, than her neighbour Canterbury. Of course all the first-class pasture-land "down south," as it was called, had been taken up long before; but we heard rumours of splendid sheep country, yet unappropriated, far back towards the west coast of Otago, just where its boundary joined Canterbury.

With our minds in this state of desire for what poor Mazzini used to denounce as "territorial aggrandisement," we paid our usual post-shearing visit to Christchurch. F—— had his agent's accounts to examine, a nice little surplus of wool-money to receive, and many other squatting interests to attend to; whilst I had to lay in chests of tea, barrels of sugar and rice, hundreds of yards of candle-wick, flower-seeds, reels of cotton, and many other miscellaneous articles. But through all our pleasant, happy little bustle ran the constant thought: "What shall we do for more country?" A day or two before the expiration of the week's leave of absence which we always gave ourselves, F—— came into my sitting-room at the hotel, flung down his hat on the table with an air of triumph, and cried, "I've heard of such a splendid run! One hundred thousand acres of beautiful sheep-country, and going for a mere song!" Now I had lived long enough in the world to discover that one sometimes danced on the wrong foot to the tune of these "mere songs," so I cautiously inquired, "Where is it?" F—— seemed a little dashed that the only question which he could not answer favourably should be the first I asked, and he replied vaguely, "Well, it is rather a long way off, but I am sure we can manage it." A little more sifting elicited the fact that this "desirable investment" stretched along the shores of Lake

Wanaka, famous for its beautiful scenery, and was to be had for what certainly seemed a ridiculously small sum;—only a few hundred pounds. "Of course it has no sheep on it," added F——; "but that is all the better. I'll burn it this year, and then turn some cattle on it, and after next shearing we'll have a good mob of sheep to draft out and stock it." He further added, that he had invited his man of business and the individual who owned this magnificent property to dine with us that evening, and that then I should hear all about it And I may truly say that I did hear about it, for my brain reeled with figures and calculations. By bedtime I was wondering if we could possibly spend the enormous fortune which would be quite certain to accrue to us in a few years if only we could make up our minds to invest the modest balance at our bankers in this tempting bargain. I remember well that I found myself wishing we were not going to be quite so rich; half our promised income would have been ample, I thought. My anxieties on that score turned out to have been, to say the least, premature.

Not to make my story too long, I may briefly say that after making due allowance for the natural exaggeration of the owner, the run on Lake Wanaka's shores seemed certainly to offer many attractions. Besides thousands of acres of beautiful sheltered sheep country, it was said to possess a magnificent bush, in which sawyers were already hard at work. Of course all this timber would become our own, and we were to make so much a year by selling it. "How about the carriage?" inquired F—— cautiously, having visions of costly bullock-drays, and teams and drivers at fabulous wages. "Oh, the lake is your highway," replied the would-be seller, airily; "you have nothing to do but lash your felled trees together, as they do in the mahogany-growing countries, and set them afloat on the lake, they will thus form a natural raft, and cost you little or nothing to get down to a good market. You know the Dunstan diggings are just at the foot of the lake, and they haven't a stick there; timber is very badly wanted in those parts, not only for fuel and building, but also for slabbing the shafts which the miners sink."

By the time the coffee was served F—— had made up his mind to buy the Lake Wanaka run; his business agent urging him strongly not to hesitate for a moment in securing such a chance. The negotiations reached thus far without the least hitch, but at this point F— —said, "Well, I'll tell you what I'll do: we will start in a day or two and go straight up to this run and look round it, and if I find it anything like so good as you both make it out, I'll buy it on the spot."

Never did that sociable little word "we" sound so delightful to my ears! "Then I am to come too," I thought to myself, but I prudently concealed from the company that I had ever had any misgivings on that point. However, the company did not concern themselves with my doubts and fears, for our two guests seemed much taken aback at this very matter-of-fact proposal of F— —'s. "That won't do at all, my dear fellow," said the owner of the run; "I am going to England by the next mail steamer, which you know sails next week, and the reason I am literally giving away my property is that I don't want any suspense or bother. Take it or leave it, just as you like. There's Wilkinson and Fairwright and a lot of others all clamouring for the refusal of it, and I've only waited to see if you really wanted it before closing with Fairwright. He is walking about with a cheque all ready filled up in his pocket, and only begging and praying me to let him have the run on my own terms. Why you might be weather-bound or kept there for a month, and what shall I do then? No, its all just as I've told you, and you can call it your own to-morrow, but I can't possibly wait for you to go and look at it." No words of mine can give any idea of the tone of scorn in which our guest pronounced these last three words; as if looking at an intended purchase was at once the meanest and most absurd thing in-the world. F— — seemed half ashamed of himself for his proposal, but still he urged that he never liked to take a leap in the dark, backing up his opinion by several world-revered adages. "That's all very fine," chimed in our precious business adviser," but this transaction can hardly be said to be in the dark; here are the plans and the Government lease and the transfer deeds, all regular and ready." With this he produced the plans, and then it was all up with us. Who does not know the peculiar smell of tracing-paper, with its suggestions of ownership? When these fresh and crackling drawings were opened before us they resembled nothing so much as a veritable paradise. There shone the lake—a brilliant patch of cobalt blue, bordered by outlines of vivid green pasture and belts of timber. Here and there, on the outskirts, we read the words, "proposed township," "building lots," "probable gold fields," "saw mills." F— — laid his hand down over a large wash of light green paint and asked," Now what sort of country is this; really and truly, you know?" "First class sheep country, I give you my word," replied the owner eagerly, "only wants to be stocked for a year or two."

Why need I go on? It was the old, old story of misplaced confidence. Neither F— — nor I could believe that our friends would wilfully over-reach us, so it was settled that the first thing next morning the money should be handed over and the Government lease transferred to us. We decided that as we were so far on the way to our new property, we would go and look at it before returning to the Malvern Hills, and the next few days were very busy ones, as we had to arrange our small domestic affairs, send up the dray, etc., etc. I felt rather anxious at the postponement of our return home, for I had left several "clutches" of eggs on the point of being hatched, and I had grave misgivings as to the care my expected ducklings and chickens would receive at the lands of my scatter-brained maid servants, to say nothing of the dangers besetting them from hawks and rats. However, small interests must give way to great ones, and F— — and I were already tasting the cares of proprietorship. Our friend, the former owner of our new property, sailed for England in the mail steamer, in high spirits, saying cordially as he shook F— —'s hand at parting, "Well youhave got your fortune cut out for you, and no mistake; I feel half sorry already to think that I've parted with that run." About two days after his departure, F— — who had registered his name at the land office as the present tenant of 100,000 acres in the Lake Wanaka district, received a polite request from official quarters to pay up the annual rent, just due, amounting to 100 pounds or so. We had effected our brilliant negotiations about a week too soon it seemed, but that was our own fault, so we had nothing to do but pay the money with as good a grace as possible. I am "free to confess" that this second cheque ran our banker's account very fine indeed, but still in those palmy days of the past this was no subject of uneasiness to a squatter. His credit was almost unlimited, and he could always raise as much money as he liked on an hypothecation of next year's wool. But we had not come to that yet. The weather was delightful; the customary week of heavy rain just after our midsummer Christmas, had cooled the air and laid the dust, besides bringing out a fresh spring-like green tint over the willows and poplars, and causing even the leaves of the gums to lose their leather-like look for a few days.

After much consultation we decided to go by coach as far as Timaru, and then trust to circumstances to decide our future means of transport. Not only were we obliged to pay a large sum for our places but our luggage was charged for by the pound, so we found it necessary to reduce our kit to the most modest dimensions, and only to take what was absolutely necessary. The journey was a long and weary one, the only variety being caused by a strong spice of danger at each river. At some streams we were transferred bodily to a large raft-like ferry boat, and so taken across. At others the passengers and luggage only were put into the boat, the lumbering coach with its leathern springs left behind, whilst the horses swam in our wake across the wide and rushing river, to be re-harnessed to another coach on the opposite shore. The Rakaia, Ashburton, and Rangitata had been crossed in this way, and we had reached the Otaio, a smaller river, when we found a new mode of transport awaiting us. A large dray with a couple of powerful horses was in readiness, and into this springless vehicle we were unceremoniously bundled. The empty coach and horses was driven over at another part of the stream. I shall never forget the jolting: the river must have been at least a quarter of a mile wide at that reach, and over its bed of boulders and rocks we bumped In the middle stretched a long strip of shingle, which seemed as smooth as turf by contrast with the first half of the river-bed. When we charged into the water again our driver removed his pipe from his mouth, looked over his shoulder and remarked, "River's come down since mornin'; best tuck up your feet, marms all." I can answer for this "marm" tucking up her feet with great agility, and not a moment too soon either, for as a light wind was blowing, a playful wave came rippling over and through the planked floor of the dray, floating all the smaller parcels about. But no one could speak, we were so jolted: it literally seemed as if our spines must come through the crown of our heads, and I expected all my teeth to tumble out.

In the midst of my fright and suffering, a laugh was jolted out of me by the absurd behaviour of one of our fellow-passengers. He was what is called a bush carpenter: i.e., a wandering carpenter, who travels from station to station, doing any little odd rough jobs wanted. This man had been working for us some time before, and had often amused me with his quaint ways. On this occasion he was on his oppressively good behaviour, and sat quite silent and solemn on the opposite ledge of the dray. But when for the second time the water came swirling through our rude conveyance with a force which threatened to upset it altogether, Dale fumbled in his pocket, as if he were seeking for a life-belt, produced an enormous pair of green goggle spectacles, which might have made part of Moses Primrose's purchases at the fair, and adjusting them on his nose as steadily as he could, said gravely, "This must be looked to!" He continued to stare at the wash of water during the remainder of our perilous and rough transit without vouchsafing any explanation of his meaning, but after we had safely landed he replaced his spectacles, first in their huge shagreen case, and next in his pocket, with an air which seemed to say, "The danger is now over: thanks to my precautions."

Timaru was reached very late, and the best accommodation at the inn placed at our disposal. Still, in those distant days there was no such thing as a private sitting room, and we had all to eat our supper in the same rough-boarded little apartment. But in all my varied wanderings in different parts of the world, when the accidents of travel have thrown me for a time among the class whom we foolishly speak of as the lower orders, I have never yet had to complain of the slightest inconvenience or disagreeableness from my fellow-travellers. On the contrary, I have always received the most chivalrous politeness at their hands, and have noticed how ready they were to forego their usual tastes and habits lest they should cause me any annoyance. I wonder whether fine gentlemen in their splendid clubs would be quite so willing to spoil the pleasure of their evening if any accident were to throw an unwelcome lady amongst them?

At all events, they could not be more self-sacrificing than my friends in fustian jackets have always proved themselves, and on this particular evening the landlord of the inn was so amazed at the orders for tea and coffee instead of the usual "nips" of spirits, that he was constrained to inquire the reason. A stalwart drover who was sitting opposite to me at the rude table, murmured from the depths of his great beard, in an oracular whisper, "The smell of speerits might'nt be agreeble like to the lady." In vain I protested that I did not mind it in the least; tea and coffee was the order of the evening, and solemn silence and good behaviour. No smoking, no songs, no conviviality of any sort. I would fain have shown my appreciation of their courtesy by talking to them; but alas, I was one vast ache all over! Although the road had been a dead level, sixteen hours of jolting and bumping had reduced me to a limp, black-and-blue creature, with out a word or a smile. Of course I retired to what was literally a pallet, and a very hard pallet too, as early as possible, but even after I had vanished behind the thin wooden partition which formed my bedroom, the greatest silence and decorum continued to reign among my fellow-travellers.

Chapter VII: "Buying a run."—continued.

Early the next morning we all breakfasted together, and then separated with most polite adieux. We sallied forth to look for a couple of riding horses. There were none to be hired, so we had to buy two good-looking nags for 45 pounds a-piece. Now-a-days the same horses would not fetch more than 10 pounds and I have been told that in Australia you can buy a horse for a shilling, but ours in New Zealand have never sunk lower than a couple of pounds, if they had any legs at all. It seemed to the horse-dealer quite a superfluous question when I timidly inquired if my horse had ever carried a lady. "No: I can't just say as he has, mum, as you see there aint no ladies in these parts for him to carry. But," he added magnanimously, "I'll try him with a blanket fust, if you're at all oneasy about him." We did not start until the next day, as we had to hunt up side-saddles, and I had to sew a few yards of grey linsey into a riding-skirt; but by the following day we were all ready, and our "swags" packed and strapped to the saddles by nine o'clock. F——'s horse looked a very nice one in every respect; mine was evidently uneasy in his mind at the strange shape of his saddle, and I was recommended to mount outside the little enclosure, on a patch of open ground, where my steed would not be able to brush me off. The moment I mounted, the "Hermit" as he was called, made for a dry ditch and tried to lie down, but a sharp cut from a stock-whip brought him out of it, and then he laid his ears well back and started for a good gallop, to endeavour to get rid of his strange rider. However, his head was turned in the right direction; there were no obstacles in the way, and before he got tired of his pace we had left Timaru a good many miles behind us. F—— looked complacently at the "Hermit," and observed, "He'll carry you very nicely, I think." I could only breathe a sincere hope that he might.

It was a beautiful day, warm but not oppressive, and delightfully calm. Our road lay at first along the sea-shore. Ever since we had left Christchurch the ground had been almost level, and the road consisted merely of a track cleared from tussocks. On our left extended the vast strip known as the Ninety-miles Beach, whilst far on our right, between us and the west coast, the Southern Alps, rose in all their might and beauty, sometimes lightly veiled by a summer haze, at others cutting our Italian-blue sky sharp and clear with their grand outlines. Our horses were a trifle too fat for good condition, and we feared to hurry them the first day, so we made an early halt at Mahiki, only a twenty miles stage; but the next day they took us on to Waitaki Ferry, past a splendid bush, and so into the heart of the hill country.

Between the ranges, beautiful fertile valleys extended; when I say fertile, I mean that the soil was excellent, and the land well-grassed. But there was no cultivation. Not a sod had ever been turned there since the creation of the world, and the whole country wore the peculiar yellow tinge caught from the tall waving tussocks, which is the prevailing feature of New Zealand scenery au naturel. Every acre had been "taken up," but as yet the runs were rather understocked. Our fourth day's ride was the longest,—fifty-five miles in all, though we halted for a couple of hours at a miserable accommodation house. Our bivouac that night was close to Lake Wanaka, at the Molyneux Ferry-house, and there I was kept awake all night by the attentions of a cat. I never saw such a ridiculous animal. Prince, for that was his name, took the greatest fancy to me, or rather to my woollen skirt I suppose, and found a linsey lap much more comfortable than the corduroy knees on which he took his usual evening nap. At all events he followed me into my room, which only boasted of a mattress, stuffed with tussock-grass by the way, on the floor. Here I should have slept very well after my long journey, if Prince would have permitted it. In vain I put him out of the window, not always very gently; he returned in five minutes, bringing a palpitating, just-caught bird or mouse, which he softly dropped on my face, and purred loudly with delight at his own gallantry. Twenty times did I strike a match that night and try to restore the victims to life; only one recovered sufficiently to be released, and Prince brought it in again, quite dead, five minutes later. I shut the little casement window, but the room became so hot and stuffy, and suspicious fumes of stale beer and tobacco began to assert their presence, so that I found myself obliged to open it again. Sometimes the victim's bones were crunched close to my ear, and I found more than one feather in my hair in the morning. Never was any one so persecuted by a cat as I was by Prince that weary night.

The next day we got to a station known as "Johnson's." It was just at the head of the lake, and as we arrived tolerably early in the forenoon we embarked, after the usual station dinner of mutton, tea, and damper, on Lake Wanaka. Alas for those treacherous blue waters! We had only a little pair-oared boat, in which I took my place as coxwain, and after pulling for a mile or two under a blazing sun, over short chopping waves, with a head-wind, we all became so deadly sea-sick that we had to turn back! As soon as we had rested and recovered, a council of war was held as to our movements, and we decided, in spite of our recent experiences, to turn our horses, who had done quite enough for the present, out on the run, and so make our way down the lake by boat. Already F— — was beginning to look anxious, for he perceived that, even after the head of the lake had been reached, the wool would cost an enormous sum to cart down to either Oamaru or Timaru, from whence alone it could be shipped.

The mile or two of the run which lay along the shore of the lake showed us frightfully rough country. A dense jungle of tussocks and thorny bushes choked up the feed, and made it impossible to drive any animals through it, even supposing that good pasturage lay beyond. Still we hoped that we might be looking at the worst portion of our purchase, and deter mined to persevere in the attempt to penetrate to the furthest end of our new property. Accordingly we hired a safe old tub of a boat which, though too heavy to pull, was warranted to sail steadily, and with a couple of men, some cold mutton, bread, tea, and sugar, started valiantly on our cruise. But the "blue, unclouded weather," in which we had hitherto basked, was at an end for the present. We had already enjoyed a longer succession of calm days than usually falls to the lot of the travellers in that windy middle island, and it was now quite time for the imprisoned "nor'-wester" to have his turn over the surface of the domain.

Accordingly the first day's sail was against a light, ominously warm head-wind, and we only made any way at all by keeping up a complicated system of tacking. The start had not been an early one, so darkness found us but little advanced on our voyage, and we passed the night in a rough shanty, on beds of fern-leaves, wrapped in our red blankets. Tired as we were, none of us could sleep much. The air was dry and parched; every now and then a sough of the rising hot gale swept through our crazy shelter without cooling us, and warned us to prepare for what was coming. Our only chance of getting on was to make an early start, for fortunately a true "nor'-wester" is somewhat of a sluggard. The skies wore their peculiar chrysoprase green tint, except towards the weather quarter, where heavy banks of lurid cloud showed that the enemy was collecting in force. Even the hour of dawn, usually so crisp and cool, brought no sense of refreshment to our languid limbs, and we embarked with the direst forebodings. A few miles further up the lake we reached an out-station hut, built by our host Mr. Johnson when he first "took up" his country and intended to push his boundary as far as this. He soon drew in his lines however on account of the rough nature of the ground. The hut was in a most picturesque spot, and although deserted, remained still in good repair. The little scrap of garden ground was a tangle of gooseberry and currant bushes among which potatoes flourished at their own sweet will.

We had only time to beach the boat, that is to say F— — and the two men did so, whilst I ran backwards and forwards with the blankets and provisions, before the hurricane was upon us. Henceforth there was no stirring out of doors until the gale had blown itself out. We dragged in some driftwood, barricaded the door, and prepared to pass the time as well as we could. Oh, the fleas in the hut! The ground was literally alive with them, and their audacity and appetite was unparalleled. Our boatmen sat tranquilly by the tiny window and played cribbage incessantly with very dirty cards and a board made out of a small bar of soap. As for me, I turned an empty box up on its end, so as to get out of the way of the fleas, and perched myself on it, finding ample occupation in defending my position from the attacks of the active little wretches. Sometimes I felt as if I must rush out into the lake and drown myself and my tormentors together. It was very bad for everybody. The poor boatmen doubtless wished to smoke, but were too polite to do anything of the sort. F— — had nothing whatever to read, except a torn piece of an old Times, at least two years old, which we had brought to wrap up some of our provisions; whilst I was still more idle and wretched. Two weary interminable days dragged, or perhaps I should say, blew, themselves along in this miserable fashion, but at sundown on the evening of the third day the wind dropped suddenly, and we did not lose a moment in darting out of our prison and embarking once more. For the first time since we started we could perceive the grandeur of the surrounding country; but grand scenery is not necessary nor indeed desirable in a sheep run. Splendid mountains ran down in steep spurs to the very shore of the enormous lake. Behind them, piled in snowy steeps, rose the distant Alps of the Antipodes; great masses of native bush made dark purple shadows among the clefts of the hills, whilst the lake rippled in and out of many a graceful bay and quiet harbour. Not a fleck or film of cloud floated between us and the serene and darkening sky; a profound, delightful calm brooded over land and water. Although there was no moon, the stars served us as lights and compass until two o'clock in the morning, by which time we had reached the head of the lake (which is thirty-five miles in length), where we landed, extemporized a tent out of the boat sail, and turned in for a refreshing flea-less sleep.

The next day was beautifully still, with a light air from the opposite point, just sufficient to cool the parched atmosphere; and we made our way along the head of the lake to a place were a couple of sawyers were at work. One of them had brought his wife with him, and her welcome to me was the most touching thing in the world. She took me entirely under her care, and would hardly let me out of her sight. I must say it was very nice to be waited on so faithfully, and I gave myself up to the unaccustomed luxury. All she required of me in exchange for her incessant toil on my behalf was "news." It did not matter of what kind, every scrap of intelligence was welcome to her, and she refused to tell me to what date her "latest advices" extended. During the three days of our stay in that clearing among the great pines of the Wanaka Bush, I gave my hostess a complete abridgment of the history of England—political, social, and moral, beginning from my earliest recollections. Then we ran over contemporary foreign affairs, dwelt minutely on every scrap of colonial news, and finally wound up with a full, true, and particular account of myself and all my relations and friends. When I paused for breath she would cease her washing and cooking on my behalf, and say entreatingly, "Go on now, do!" until I felt quite desperate.

All this time whilst I was being "interviewed" nearly to death, F—— employed himself in making excursions to different parts of the run. One of the sawyers lent him a miserable half-starved little pony; and he penetrated to another sawyer's hut, seven miles distant up the Matukituki river. But no matter whether he turned his steps to north or south, east or west, he met with the same disheartening report. There was the ground indeed, but it was perfectly useless. Not only was there was no pasturage, but if there had been, the nature of the country would have rendered it valueless, on account of the way it was overgrown. It would be tedious to explain more minutely why this was the case. Sufficient must it be to say that whilst F—— was only too anxious to keep his eyes shut as to the ground he had alighted on after his leap in the dark, and the sawyers were equally anxious to induce settlers to come there, and so bring a market for their labour close to their hand nothing could make our purchase appear anything except a dead loss. As for the plans, they were purely imaginary. The blue lake was about the only part true to nature; and even that should have had a foot-note to state that it was generally lashed into high, unnavigable waves, by a chronic nor'-wester.

No: there was nothing for it but to go home again to the little run which had seemed such a mere paddock in our eyes, whilst we indulged in castle-building over 100,000 acres of country. It was of no use lingering amid such disappointment and discomfort; besides which my listener, the sawyer's wife, had turned her husband and herself out of their hut, and were sleeping under a red blanket tent. Poor woman, she was most anxious to get away; and the lovely sylvan scene, with the tall trees standing like sentinels over their prostrate brethren, the wealth of beauteous greenery, springing through fronds of fern and ground creepers, the bright-winged flight of paroquets and other bush birds, even the vast expanse of the lake which stretched almost from their threshold for so many miles, all would have been gladly exchanged for a dusty high street in any country town-ship. Her last words were, "Can't you send me a paper or hany thing printed, mam?" I faithfully promised to do my best, and carried out my share of the bargain by despatching to her a large packet of miscellaneous periodicals and newspapers; but whether she ever received them is more than I can say.

We were afraid of lingering too long, lest another nor'-wester should become due; and we therefore started as soon as F— — had decided that it was of no use exploring our wretched purchase any further. We had a stiff breeze from the north-west all the way down the lake; but as it was right a-stern it helped us along to such good purpose, that one day's sailing before it brought us back to Mr. Johnson's homestead and comparative civilization. The little parlour and the tiny bed-room beyond, into which I could only get access by climbing through a window (for the architect had forgotten to put a door), appeared like apartments in a spacious palace, so great was the contrast between their snug comfort and the desolate misery of our hut life. Of course nothing else was talked of except our disappointment at our new run; and although Mr. Johnson had indulged in forebodings, which were only too literally fulfilled, he had the good taste never to remind us of his prophecies.

"Of all the forms of human woe,
Defend me from that dread, 'I told you so.'"

After a day's halt and rest we mounted our much refreshed horses, and set our faces straight across country for Dunedin. This is very easy to write, but it was not quite so easy to do. We could only ride for the first fifty-two miles, which we accomplished in two days. These stages brought us to the foot of the Dunstan Range, and near the gold-diggings of that name. I would fain have turned aside to see them, but we had not time. However, we felt the auriferous influence of the locality; for a perfect stranger came up to us, whilst we were baiting at another place, called the Kaiwarara diggings, and offered to buy our horses from us for 30 pounds each, and also to purchase our saddles and bridles at a fair price. This was exactly what we wanted, as we had intended to sell them at Dunedin; and I was no ways disinclined to part with the Hermit; who retained the sulky, misanthropical temper which had earned him his name. He was now pronounced "fit to carry a lady," and purchased to be sold again at the diggings. Whether there were any ladies there or not I cannot tell. Of course, before parting with our nags we ascertained that the ubiquitous

"Cobb's coach" started from our resting place for Dunedin next day, and we made the rest of our journey in one of that well-known line. Its leathern springs, whilst not so liable to break by sudden jolts, impart a swinging rocking motion to the body of the vehicle, which is most disagreeable; but rough and rude as they are, they deserve to be looked upon with respect as the pioneers of civilization. All over America, Australia, and now New Zealand, the moment half-a-dozen passengers are forthcoming, that moment the enterprising firm starts a coach, and the vehicle runs until it is ousted by a railway. All previous tracks which I had journeyed over seemed smooth turnpike roads, compared to that terrible tussocky track which led to Dunedin.

But that bright little town was reached at last, the hotel welcomed us, tired and bruised travellers that we were, and next evening we started in the Geelong for Port Lyttleton. This little coasting steamer seemed to touch at every hamlet along the coast, and after each pause I had to begin afresh my agonies of sea-sickness. There was no such thing as getting one's sea-legs; for we were seldom more than a few hours outside, and had no chance of getting used to the horrible motion. Timaru was reached next day, but we had suffered so frightfully during the night from a chopping sea and an open roadstead, that we went on shore, and entrusted ourselves once more to the old coach. It seemed better to endure the miseries we knew of, than to make experiments in wretchedness. So we went through the old jolting and jumbling until we were dropped at an accommodation house, fifteen miles from Christchurch, where we slept that night, and at daylight despatched a messenger to the next station for our own horses. He had only thirty-five miles to ride, and about mid-day we started to meet him on hired horses, which we were very glad to exchange for better nags a stage further on.

And so we rode quietly home in the gloaming, winding up the lovely, tranquil valley, at whose head stood our own snug little homestead. At first we were so glad to be safely at hone again that we scarcely gave a thought to our fruitless enterprise; but as our bruised bodies became rested and restored, our hearts began to ache when we thought of the money we had so rashly flung away in BUYING A RUN.

Chapter VIII: Looking for a congregation.

It is to be hoped and expected that such a good understanding has been established between my readers and myself by this time, that they will not find the general title of these papers unsuitable to the heading of this particular chapter. Indeed, I may truly say, that, looking back upon the many happy memories of my three years life in that lovely and beloved Middle Island, no pleasures stand out more vividly than my evening rides up winding gullies or across low hill-ranges in search of a shepherd's hut, or a cockatoo's nest. A peculiar brightness seems to rest on those sun-lit peaks of memory's landscape; and it is but fitting that it should be so, for other excursions or expeditions used to be undertaken merely for business or pleasure, but these delicious wanderings were in search of scattered dwellings whose lonely inhabitants—far removed from Church privileges for many a long year past—might be bidden, nay, entreated, to come to us on Sunday afternoons, and attend the Service we held at home weekly.

And here I feel constrained to say a word to those whose eyes may haply rest on my pages, and who may find themselves in the coming years in perhaps the same position as I did a short time ago. A new comer to a new country is sure to be discouraged if he or she (particularly she, I fancy) should attempt to revive or introduce any custom which has been neglected or overlooked. This is especially the case with religious observances. At every turn one is met by disheartening warnings. "Oh, the people here are very different to those in the old country; they would look upon it as impertinence if you suggested they should come to church." "You will find a few may come just at first, and then when the novelty wears off and they have seen all the pretty things in your drawing room, not a soul will ever come near the place."

"If even the men don't say something very free and easy to you when you invite them to your house on Sunday afternoons, you may depend upon it that after two or three weeks you will not know how to keep them in order."

Such, and many more, were the discouraging remarks made when I consulted my neighbours about my plan for collecting the shepherds from the surrounding runs, and holding a Church of England Service every Sunday afternoon at our own little homestead. To my mind, the distances seemed the greatest obstacle, as many of the men I wanted to reach lived twenty-five or even thirty miles away, with very rough country between. I had no fear of impertinence, for it is unknown to me, and seldom comes, I fancy, unprovoked; whilst with regard to the novelty wearing off and the men ceasing to attend, that must be left in God's hands. We could only endeavour to plant the good seed, and trust to Him to give the increase. It was a great comfort to me in those early days that F— —, who had been many years in the colony, never joined in the disheartening prophecies I have alluded to. Although as naturally averse to reading aloud before strangers as a man who had lived a solitary life would be sure to be, he promised at once, with a good grace, to read the Evening Service and a sermon afterwards, and thus smoothed one difficulty over directly. His advice to me was precisely what I would fain repeat: "Try, by all means: if you fail you will at least feel you have made the attempt." May all who try succeed, as we did! I believe firmly they will, for it is an undertaking on which God's blessing is sure to rest, and there are no such fertilizing dews as those which fall from heaven. The mists arising from earth are only miasmic vapours after all!

But I fear to linger too long on the end, instead of telling you about the means.

It was May when we were fairly settled in our new home at the head of a hill-encircled valley. With us that month answers to your November, but fogs are unknown in that breezy Middle Island, and my first winter in Canterbury was a beautiful season, heralded in by an exquisite autumn. How crisp the mornings and evenings were, with ever so light a film of hoar frost, making a splendid sparkle on every blade of waving tussock-grass! Then in the middle of the day the delicious warmth of the sun tempted one to linger all day in the open air, and I never wearied of gazing at the strange purple shadows cast by a passing cloud; or up, beyond the floating vapourous wreath, to the heaven of brilliant blue which smiled upon us. And yet, when I come to think of it, I don't know that I had much time to spare for glancing at either hills or skies, for we were just settling ourselves in a new place, and no one knows what that means unless they have tried it, fifty miles away from the nearest shop. The yeast alone was a perpetual anxiety to me,—it would not keep beyond a certain time, and had a tendency to explode its confining bottles in the middle of the night, so it became necessary to make it in smaller quantities every ten days or so. If by any chance I forgot to remind my scatter-brained damsels to replenish the yeast bottles, they used up the last drop, and then would come smilingly to me with the remark, "There aint not a drop o' yeast, about, anywhere, mum." This entailed flap-jacks, or scones, or soda bread, or some indigestible compound for at least three days, as it was of no use attempting to make proper bread until the yeast had worked. Then the well needed to be deepened, a kitchen garden had to be made, shelter to be provided for the fowls and pigs; a shed to be put up for coals; a thousand things which entailed thought and trouble, had to be done.

It is true these rough jobs were not exactly in my line, but indoors I was just as busy trying to make big things fit into little spaces and vice versa. We could not afford to take things coolly and do a little every day, for at that time of year an hour's change in the wind might have brought a heavy fall of snow, or a sharp frost, or a; deluge of rain down upon the uncovered and defenceless heads of our live stock. The poor dear sheep, the source of our income, were after all the least well-cared for creatures on the Station. A well grassed and watered run, with sunny vallies for winter feeding, and green hills for summer pasturage, had been provided by antipodean Nature for them, and to these advantages we only added some twenty or twenty-five miles of wire fencing, and then they were left to themselves, with a couple of shepherds to look after fifteen thousand sheep all the year round.

But yet, busy as we were, we found time to look up a congregation. The very first Sunday afternoon, whilst we were still in the midst of a chaos of chips and big boxes and straw and empty china-barrels, our own shepherds came over, by invitation, and the only very near neighbours we had—a Scotch head-shepherd and his charming young wife,—and we held a Service in the half-furnished drawing room. After it was ended we had a long talk with the men, and they confessed that they had enjoyed it very much, and would like to come regularly. When questioned as to the feasibility of inducing others to join, they said that it might be suggested to more than one distant, lonely hill-shepherd, but his uncontrollable shyness would probably prevent his attendance.

"Jim Salter, and Joe Bennett, and a lot more on 'em, would be glad enow to come, if so be they could feel as how they was truly wellcombe," said our shepherd, Pepper, who prided himself on the elegance and correctness of his phraseology. He added, after a reflective pause, turning bashfully away, "If so be as the lady would just look round and give 'em a call, they'd be to be persuaded belike."

So the scheme was Pepper's after all, you see. But this "looking round," to which he alluded so airily, meant scrambling rides, varying from ten to twenty-eight miles in length, over break-neck country, and this on the slender chance of finding the men in-doors. Now a New Zealand shepherd almost lives out on the hills, so the prospect of finding any of our congregation at home was slight indeed. However, as I said before, F— — stood by me, and although we neither of us could well spare the time, we agreed to devote two afternoons every week, so long as the fine open autumn weather, lasted, to making excursions in search of back-country huts. There are no roads or finger posts or guides of any sort in those distant places. When we inquired what was the name of "Mills" shepherd (the masters are always plain Smith or Jones, and the shepherds Mr. — —, in the colonies) the answer was generally very vague. "Wiry Bill, we mostly calls 'im; but I think I've heerd say his rightful name was Mr. Pellet, mum. He's a little chap, as strong as the 'ouse," explained Pepper, who was an incorrigible cockney, "and he lives over there," pointing with his thumb to a mountain range behind us. "He's in one of them blind gullies. You go along the gorge of the river till you come to a saddle all over fern, and you drop down that, and follow the best o' three or four tracts till you come to a swamp."

Here Pepper paused, in consideration of my face of horror; for if there was one thing I dreaded more than another in those early days, it was a swamp. Steep hill sides, wide creeks, honey-combed flats, all came in, the day's ride,—but a swamp! Ugh! the horrible treacherous thing, so green and innocent looking, with here and there a quicksand or a peaty morass, in which, without a moment's warning, your horse sank up to his withers! It was dreadful, and when we came to such a place Helen used to stop dead short, prick her pretty ears well forward, and, trembling with fear and excitement, put her nose close to the ground, smelling every inch, before she would place her fore foot down on it, jumping off it like a goat if it proved insecure. Generally she crossed a swamp, by a series of bounds in and out of flax bushes; and hopeless indeed would a morass be without those green cities of refuge!

Horrible as a large swamp is however to a timid horsewoman, it is dear to the heart of a cockatoo. He gladly buys a freehold of fifty acres in the midst of one, burns it, makes a sod fence, sown with gorse seed a-top, all round his section, drains it in a rough and ready fashion, and then the splendid fertile soil which has been waiting for so many thousand years, "brings forth fruit abundantly." Such enormous fields of wheat and oats and barley as you come upon sometimes,—with, alas, never a market near enough to enable the plenteous crop to return sevenfold into its master's bosom!

I shall not inflict upon you a description of all our rides in search of members for our congregation. Two, in widely differing directions, will serve as specimens of such excursions. In consideration of my new-chumishness, F—— selected a comparatively easy track for our first ride. And yet, "bad was the best," might surely be said of that breakneck path. What would an English horse, or an English lady say, to riding for miles over a slippery winding ledge on a rocky hill side, where a wall of solid mountain rose up perpendicularly on the right hand, and on the left a very respectable sized river hurried over its boulders far beneath the aerial path; yet this was comparatively a safe track, and presented but one serious obstacle, over which I was ruthlessly taken. It is perhaps needless to say we were riding in single file, and equally unnecessary to state that I was the last; for certainly we should never have made much progress otherwise. Helen, my bay mare, would follow her stable companion, on which F—— was mounted, so that was the way we got on at all.

A sudden sharp turn showed me what appeared to be a low stone wall running own the spur of the mountain, right across our track, and I had already begun to disquiet myself about the possibility of turning back on such a narrow ledge, when I saw F— —'s powerful black horse, with his ears well forward, and his reins, lying loose on his neck, make a sort of rush at the obstacle, climb up it as a cat would, stand for an instant, exactly like a performing goat, with all four legs drawn closely together under him, and then with a spring disappear on the other side. "This wall", I thought, "must be but loosely built, for Leo has displaced some of the stones from its coping." Helen, pretty dear, hurried after her friend and leader; and before I had time to realize what she was going to do, she was balancing herself on the crumbling summit of this stone wall (which was only the freak of a landslip), and as it proved impossible to remain there, perched like a bird on a very insecure branch, nothing remained except to gather herself well together and jump off. But what a jump! the ground fell sheer away at the foot of the wall, and left a chasm many feet wide, which the horse could not see until it had climbed to the top of the wall, and as turning back was out of the question, the only alternative was to give a vigorous bound on to the narrow ledge beyond. Terrified as I felt, I luckily refrained from jerking Helen's head, or attempting to guide her in any way. The only chance of safety over New Zealand tracks, or New Zealand creeks, is to leave your horse entirely to itself. I have seen men who were reckoned good riders in England, get the most ignominious tumbles from a disregard of this advice. An up-country horse knows perfectly well the only sound spots in a swamp; or the only sound part of a creek's banks. If his rider persists in taking him over the latter, where he himself thinks it narrowest and safest, he is pretty sure to find the earth rotten and crumbling, and to pay for his obstinacy by a wetting; whilst in the case of a swamp the consequences are even more serious, and the horse often gets badly strained in floundering out of a quagmire.

But it was not all danger and difficulty, and the many varieties of scene in the course of a long ride constituted some of its chief charms. At first, perhaps, after we had left our own fair valley behind, the track would wind through the gorge of a river, with lofty mountains rising sheer up from the water side. All here was sad and grey, and very solemn in its eternal silence, only made more intense by the ceaseless monotonous roar of the ever-rushing water. Then we would emerge on acres and acres of softly rolling downs, higher than the hillocks we call by that name at home, but still marvellously beautiful in their swelling curves all folding so softly into each other, and dotted with mobs of sheep, making pastoral music to a flock-owner's ear. Over this sort of ground we could canter gaily along, with "Hector," F— —'s pet colley, keeping close to the heels of his master's horse,—for it is the worst of bad manners in a colley to look at a neighbour's sheep. The etiquette in passing through a strange run is for the dog to go on the off side of his master's horse, so that the sheep shall not even see him; and this piece of courtly politeness Hector always practised of his own accord.

A wire fence always proved a very tiresome obstacle, for horses have a great dread of them, and will not be induced to jump them on any account. If we could find out where the gate was, well and good; but as it might be half a dozen miles off, on one side or the other, we seldom lost time or patience in seeking it. When there was no help for it, and such a fence had to be crossed, the proceedings were, always the same. F— —dismounted, and unfastened one of his stirrup leathers; with this he strapped the wires as firmly as possible together, but if the fence had been lately fresh-strained, it was sometimes a difficult task. Still he generally made one spot lower than the rest, and over this he proceeded to adjust his coat very carefully; he then vaulted lightly over himself, and calling upon me to aid by sundry flicks on Leo's flank, the horse would be induced to jump over it. This was always a work of time and trouble, for Leo hated doing it, and would rather have leaped the widest winter creek, than jumped the lowest coat-covered wire fence. Helen had to jump with me on her back, and without any friendly whip to urge her, but except once, when she caught her hind leg in the sleeve of the coat which was hanging over the fence, and tore it completely out, she got over very well. Upon that occasion F— — had to carry his sleeve in his pocket until we reached the neat little out-station hut, where Jim Salter lived, and where we were pretty sure to find a housewife, for shepherds are as handy as sailors with a needle and thread.

I shall always believe that some bird of the air had "carried the matter" to Salter, because not only was he at home, and in his Sunday clothes, but he had made a cake the evening before, and that was a very suspicious circumstance. However we pretended not to imagine that we were expected, and Jim pretended with equal success to be much surprised at our visit, so both sides were satisfied. Nothing could be neater than the inside of the little hut; its cob walls papered with, old Illustrated London News,—not only pictures but letter-press,—its tiny window as clean as possible, a new sheep-skin rug laid down before the open fireplace, where a bright wood fire was sputtering and cracking cheerily, and the inevitable kettle suspended from a hook half-way up the low chimney. Outside, the dog-kennels had been newly thatched with tohi grass, the garden weeded and freshly dug, the chopping-block and camp-oven as clean as scrubbing could make them. It was too late in the year for fruit, but Salter's currant, raspberry, and gooseberry bushes gave us a good idea of how well he must have fared in the summer. The fowls were just devouring the last of the green-pea shoots, and the potatoes had been blackened by our first frosts.

It was all very nice and trim and comfortable, except the loneliness; that must have been simply awful. It is difficult to realise how completely cut off from the society of his kind a New Zealand up-country shepherd is, especially at an out-station like this. Once in every three months he goes down to the homestead, borrows the pack horse, and leads it up to his hut, with a quarter's rations of flour, tea, sugar and salt; of course he provides himself with mutton and firewood, and his simple wants are thus supplied. After shearing, about January, his wages are paid, varying from 75 pounds to 100 pounds a year, according to the locality, and then he gets a week's leave to go down to the nearest town. If he be a prudent steady man, as our friend Salter was, he puts his money in the bank, or lends it out on a freehold mortgage at ten per cent., only deducting a few pounds from his capital for a suit of clothes, a couple of pair of Cookham boots for hill walking, and above all, some new books.

Without any exception, the shepherds I came across in New Zealand were all passionately fond of reading; and they were also well-informed men, who often expressed themselves in excellent, through superfine, language.

Their libraries chiefly consisted of yellow-covered novels, and out of my visits in search of a congregation grew a scheme for a book-club to supply something better in the way of literature, which was afterwards most successfully carried out. But of this I need not speak here, for we are still seated inside Salter's hut,—so small in its dimensions that it could hardly have held another guest. Womanlike, my eyes were everywhere, and I presently spied out an empty bottle, labelled "Worcestershire Sauce."

"Dear me, Salter," I cried, "I had no idea you were so grand as to have sauces up here: why we hardly ever use them." "Well, mum," replied Salter, bashfully, and stroking his long black beard to gain time to select the grandest words he could think of, "it is hardly to be regarded in the light of happetite, that there bottle, it is more in the nature of remedies." Then, seeing that I still looked mystified, he added, "You see, mum, although we gets our 'elth uncommon well in these salubrious mountings, still a drop of physic is often handy-like, and in a general way I always purchase myself a box of Holloway's Pills (of which you do get such a lot for your money), and also a bottle of pain-killer; but last shearing they was out o' pain-killer, they said, so they put me up a bottle o' Cain pepper, and likewise that 'ere condiment, which was werry efficacious, 'specially towards the end o' the bottle!" "And do you really mean to say you drank it, Salter?" I inquired with horror.

"Certainly I do, mum, whenever I felt out o' sorts. It always took my mind off the loneliness, and cheered me up wonderful, especial if I hadded a little red pepper to it," said Salter, getting up from his log of wood and making me a low bow. All this time F—— and I were seated amicably side by side on poor Salter's red blanket-covered "bunk," or wooden bedstead, made of empty flour-sacks nailed between rough poles, and other sacks filled with tussock grass for a mattress and pillow.

The word loneliness gave me a good opening to broach the subject of our Sunday gatherings, and my suspicions of Jim's having been told of our visit were confirmed by the alacrity with which he said, "I have much pleasure in accepting your kind invitation, mum, if so be as I am not intruding."

"No, indeed Salter," F—— said; "you'd be very welcome, and you could always turn Judy into the paddock whilst we were having service."

Now if there was one thing dearer to Salter's heart than another, it was his little roan mare Judy: her excellent condition, and jaunty little hog-mane and tail, testified to her master's loving care. So it was all happily settled, and after paying a most unfashionably long visit to the lonely man, we rode away with many a farewell nod and smile. I may say here that Salter was one of the most regular of our congregation for more than two years, besides being a member of the book club. In time, its more sensible volumes utterly displaced the yellow paper rubbish in his but library, and I never can forget the poor man's emotion when he came to bid me good-bye.

At my request he made the rough little pen and ink sketches which are here given, and as he held my offered hand (not knowing quite what else to do with it) when I took leave of him after our last home-service, when my face was set towards England, he could not say a word. The great burly creature's heart must have been nearly as big as his body, and he seemed hardly to know that large tears were rolling down his sunburnt face and losing themselves in his bushy beard. I tried to be cheerful myself, but he kept repeating, "It is only natural you should be glad to go, yet it is very rough upon us." In vain I assured him I was not at all glad to go, — very, very sorry, in fact: all he would say was, "To England, home and beauty, in course any one would be pleased to return." I can't tell you what he meant, and he had no voice to waste on explanations; I only give poor dear Jim's valedictory sentences as they fell from his white and trembling lips.

Very different was Ned Palmer, the most diminutive and wiry of hill shepherds, with a tongue which seemed never tired, and a good humoured smile for every one. Ned used to try my gravity sorely by stepping up to me half a dozen times during the service, to find his place for him in his Prayer-book, and always saying aloud, "Thank you kindly, m'm."

Chapter IX: Another shepherd's hut.

To get to Ned's hut—which was not nearly so trim or comfortable as Salter's, and stood out in the midst of a vast plain covered with waving yellow tussocks,—we had to cross a low range of hills, and pick our way through nearly a mile of swampy ground on the other side. The sure-footed horses zig-zagged their way up the steep hill-side with astonishing ease, availing themselves here and there of a sheep track, for sheep are the best engineers in the world, and always hit off the safest and easiest line of country. I did not feel nervous going up the hill, although we must have appeared, had there been any one to look at us, more like flies on a wall than a couple of people on horse back, but when we came to the ridge and looked down on the descent beneath us, my heart fairly gave way.

Not a blade of grass, or a leaf of a shrub, was to be seen on all the steep slope, or rather precipice, for there was very little slope about it; nothing but grey loose shingle, which the first hoof-fall of the leading horse invariably sent slipping and sliding, in a perfect avalanche of rubble, down into the soft bright green morass beneath. Of all the bad "tracks" I encountered in my primitive rides, I really believe I suffered more real terror and anguish on that particular hill-side than on any other. My companion's conduct too, used to be heartless in the extreme. He let the reins fall loosely on his horse's neck, merely holding their extreme ends, settled himself comfortably in his saddle, leaning well back, and turning round laughingly to me, observed, "Aren't you coming?" "Oh, not there," I cried in true melo-dramatic tones of horror; but it was all in vain, F—— merely remarked "You have nothing to do but fancy you are sitting in an arm-chair at home, you are quite as safe." "What nonsense," I gasped. "I only wish I was at home: never, never will I come out riding again." All this time the leading horse was slowly and carefully edging himself down hill a few steps to the right, then a few to the left, just as he thought best, displacing tons of loose stone and even small rocks at every movement. Helen, nothing daunted, was eager to follow, and although she quivered with excitement at the noise, echoed back from the opposite hills, lost no time in preparing to descend. Her first movement sent such showers of rubble down upon F— — and his horse, that I really thought the latter would have been knocked off his legs. "If you could keep a little more to the right, so as to send the stones clear of me, I should be very grateful," shouted F— —, who was actually near the bottom of the hill already, so sharp had been the angles of his horse's descent. I felt afraid of attempting to guide Helen, lest the least check should send us both head over heels into the quagmire below, and yet it seemed dreadful to cause the death of one's husband by rolling down cart loads of stones upon him. It could not have been more than five minutes before Helen and I stood side by side with Leo, on the only bit of firm ground at the edge of the morass. I believe I was as white as my pocket handkerchief; and if fright could turn a person's hair grey, I had been sufficiently alarmed to make myself eligible for any quantity of walnut pomade.

Fortunately the summer had proved rather a dry one, and the swamp was not so wet as it would have been after a heavy rain-fall. The horses stepped carefully from flax bushes to "nigger heads" (as the very old blackened grass stumps are called), resting hardly a moment anywhere, and avoiding all the most seductive looking spots. I thought my companion must have gone suddenly mad, when, a hawk rising up almost from beneath our horses' feet, he flung himself off his saddle and cried out, "A late hawk's nest, I declare!" And so it proved, for a little searching in a sheltered and tolerably dry spot revealed a couple of eggs, precisely like hens' eggs, until broken, when their delicate pale green inner membrane betrayed their dangerous origin. It is chiefly owing to this practice of laying in swamps that the various kinds of hawk increase and thrive as they do, for if it were possible to get at them, the shepherds would soon exterminate the sworn foe of their chickens and pigeons. They are also the great drawback to the introduction of pheasants and partridges, for the young birds have not a chance in the open against even a sparrow-hawk.

Although it is a digression, I must tell you here how, one beautiful early winter's day, I was standing in the verandah at my own home, when one of our pigeons, chased by a hawk, flew right into my face and its pursuer was so close and so heated by the chase, that it flung itself also with great violence against my head, with a scream of rage and triumph, hurting me a good deal as it dug its cruel, armed heel into my cheek. The pigeon had fluttered, stunned and exhausted to the ground, and, quick as lightning I stooped to pick it up; so great had been the impetus of the hawk's final charge that he had never perceived his victim had escaped him. The cunning of these birds must be seen to be believed. I have often watched a wary old hawk perched most impudently on the stock-yard rails, waiting until a rash chicken or duckling should, in spite of its mother's warning clucks of terror, insist on coming out from under her sheltering wings. If I took an umbrella, or a croquet mallet, or a walking stick, and went out, the bird would remain quite unmoved, even if I held my weapon pointed gun-wise towards him. But let anyone take a real gun and hold it ever so well hidden behind their back, and emerge ever so cautiously from the shelter of the shrubs, my fine gentleman was off directly, mounting out of sight with a few strokes of his powerful wings, and uttering a shriek of derision as he departed. Nothing is so rare as a successful shot at a hawk.

We consoled ourselves however on this occasion, by reflecting that we had annihilated two young hawks before they had commenced their lives of rapine and robbery, and rode on our way rejoicing, to find Ned Palmer sitting outside his but door on a log of drift wood, making, candles. In the more primitive days of the settlement, the early settlers must have been as badly off for light, during the long dark winter evenings, as are even now the poorer inhabitants of Greenland or of Iceland, for their sole substitute for candles consisted of a pannikin half filled with melted tallow, in which a piece of cork and an apology for a wick floated. But by my time all this had long been past and over, and even a back-country shepherd had a nice tin mould in which he could make a dozen candles of the purest tallow at a time.

Ned was just running a slender piece of wood through the loops of his twisted cotton wicks, so as to keep them above the rim of the mould, and the strong odour of melted mutton fat was tainting the lovely fresh air. But New Zealand run-holders have often to put up with queer smells as well as sights and sounds, therefore we only complimented Ned on being provident enough to make a good stock of candles before-hand, for home consumption, during the coming dark days. After we had dismounted and hobbled our horses with the stirrup leathers, so that they could move about and nibble the sweet blue grass growing under each sheltering tussock, I sat down on a large stone near, and began to tell Ned how often I had watched the negroes in Jamaica making candles after a similar fashion, only they use the wax from the wild bee nests instead of tallow, which was a rare and scarce thing in that part of the world. I described to him the thick orange-coloured wax candles which used to be the delight of my childhood, giving out a peculiar perfuming odour after they had been burning for an hour or two,—an odour made up of honey and the scent of heavy tropic flowers.

Ned listened to my little story with much politeness, and then, feeling it incumbent on him to contribute to the conversation, remarked, "I never makes candles ma'am without I thinks of frost-bites."

"How is that, Palmer?" I asked, laughingly. "What in the world have they to do with each other?"

"Well, ma'am, you see it was just in this way. It was afore I come here, which is quite a lively, sociable place compared to Dodson's back country out-station, at the foot o' those there ranges beyond. I give you my word, ma'am, it used always to make me feel as if I was dead, and living in a lonely eternity. Them clear, bright-blueglassers (glaciers, he meant, I presume) was awful lonesome, and as for a human being they never come a-nigh the place. Well as I was saying, ma'am, one day I finds I had run out o' candles, and as the long dark evenings (for it was the height o' winter) was bad enough, even with a dip burning, to show me old Spot's face for company, I set to work, hot haste, to make some more. It was bitter, biting cold, you bet, ma'am; and I was hard at work—just after I had had my bit o' breakfast, before I went out for to look round my boundary—melting and making my dips, so that they might be fine and hard for night. I ought praps to mention that Spot used to get so close to the fire-place, that as often as not, I dropped a mossel of the hot grease on the dog; and if it touched a thin place in his coat, he would jump up howling. Well, ma'am, I was pouring a pannikin full o' biling tallow into the mould, when poor old Spot he gives a sudden howl and yell, and runs to the door. I paid no attention to him at the time, for I was so busy; but he went on leaping up and howling as if he had gone mad. As soon as I could put down the pannikin out o' my hand, I went to the door meaning to open it and,—sorry am I to say it,—kick the poor beast out for making such a row about a drop o' hot grease. But the dog turned his face round on me, and gave me a look as much as to say, 'Make haste, do; there's a good chap: I ought to be outside there.' And what with the sense shinin' in his eyes, and a curious kind o' sound outside, I takes down the bar (for the door wouldn't stay shut otherwise), and looks out. Never until my dyin' day, and not even then, I expect, shall I forget what the dog and I saw lying on the ground, which was all white and hard with frost, the sun not having got over the East range yet. The dog he had more sense and a deal more pluck than I had, for he knows there aint a moment to be lost; and he runs up to the flat, tumbled-down heap o' clothes, gets on its back (for no face could I see), so as to be doing something, and not losing time, and begins licking. Not very far off there was a lean horse standing, but he didn't seem to like to come through the slip-rail o' the paddock fence.

"In coorse I couldn't stand gaping there all day, so I went and stooped down to the man, who was lying flat on his face, with his arms straight out. He wasn't sensibleless (Palmer's favourite word for senseless), for he opened his eyes, and said, "For God's sake, mate, take me in." "So I will, mate," I makes reply "and welcome you are. Can you get on your legs, think you?" With that he groans awful, and says, "My legs is friz." Well, I looks at his legs, and sees he was dressed in what had been good moleskins, and high jack riding-boots, coming up to his knees; but sure enough they was as hard as a board, and actially, if you'll believe me, ma'am, there was a rim o' solid hice round the tops of his boots. As for standing, he couldn't do it: his legs was no more use to him than they was to me, and he was a tall, high fellow besides. Cold as it was, I felt hot enough by the time I had lugged that poor man inside my place, and got him up on my bunk. He could speak, though his voice was weak as weak could be, and he helped me as well as he could by catching hold with his arms, but his legs was stone dead. I had to get the tommy (anglice-tomahawk), and chop his boots off, and that's the gospel truth, ma'am. I broke my knife, first try, and the axe was too big. He told me, poor fellow, that two days before, as he was returning from prospecting up towards the back ranges, his horse got away, and he couldn't catch him. No: he tried with all his might and main, for in his swag, which was strapped to the D's of his saddle, was not only his blanket, but his baccy, and tea, and damper, and a glass o' grog. The curious thing, too, was that the horse didn't bolt right away, as they generally do: he jest walked a-head, knowing his master was bound to follow wherever he led, for in coorse he had hopes to catch him every moment. That ere brute, he never laid down nor rested,—jest kep slowly moving on, as if he was a Lunnon street-boy, with a bobby at his heels. Through creeks and rivers and swamps he led that poor fellow. His boots got chuck full o' cold water, and when the sun went down it friz into solid hice; and that misfortnit man he felt his legs—which was his life, you see, ma'am—gradially dyin' under him. Yet he was a well-plucked one, if ever there was such a party on this airth. He told me he had took five mortial hours to come the last mile, the horse walkin' slowly afore him, and guiding him like. And how do you think he did it, with two pillars of hice for legs? Why he lifted up just one leg and then the other with both his hands, and put them afore him, and took his steps that way."

Here honest Ned, his eyes glistening, and his ugly little face glowing with emotion through its coating of sunburn, paused, as if he did not like to go on.

I was more touched and interested than I could avoid showing, and cried, "Oh, do tell me, Palmer, what became of the poor fellow! Did he die?"

Ned cleared his throat, and moved so as to get between me and the light from the door, as he said huskily, "He came very nigh to it, ma'am. I never did set eyes on such a decent patient chap as that man was. I did the very wust thing I could a' done, the town doctors told me, for I brought him into the hut, instead o' keeping him outdoors and rubbing his poor black legs with snow. 'Stead o' that, I wrapped him up warm in my own blankets, after I had chipped his boots and the hice off of 'em, and I made up a roarin' fire. Good Lord, how the poor fellow groaned when he begun to get warm! I gave him a pannikin full o' hot tea, with a drop o' grog in it, and that seemed to make him awful bad. At last he said, with the sweat from sheer agony pouring down his face, "Look here, matey: couldn't you hump me out in the snow again? for it aint nigh so bad to bear it cold as it is to bear it hot." Not a bad word did he say, ma'am, and he tried not to give in more nor he could help; but he was clean druv wild with the hanguish in his legs.

"Presently I remembers, quite sudden like, that a bush doctor, name of Tomkins, was likely to be round by Simmons, cos' o' his missus. So I got on my 'oss in a minnit, and I rides off and fetches him, for sure enough he was there; and though Simmons' missis wasn't to say over her troubles, she spoke up from behind the curtain of red blanket she had put up in her tidy little hut, and bade old Tomkins go with me. May God bless her and hers for that same, say I! Well, ma'am, when Tomkins come back with me and saw the poor fellow (he was fair shoutin' with the pain in his legs by then), he said nothin' could be done. "They'll mortify by morrow mornin'," says he, "and then he'll die easy." So with that he goes back with the first light next day, to Simmons. Sure enough, the poor fellow did get a bit easier next day, and I felt clear mad to think he was goin' to die before my very eyes. "Not if I can help it!" I cries, quite savage like. But he only smiled a patient smile, and said, "God's will be done, mate. He knows best, and I aint in any pain to speak of, now."

"By and bye I hears a rumbling and a creaking, and cracking of whips; and when I looks out, what do I see but the bullock-dray from Simmons' coming up the flat. It was the only thing on wheels within forty mile, and Simmons had brought it his own self to see if we couldn't manage to get the poor fellow down to the nighest town. I won't make my yarn no longer than I can help, ma'am, so I'll only mention that we made a lot o' the strongest mutton broth you ever tasted; we slung a hammock of red blankets in the dray, and we got the poor fellow down by evening to a gentleman's station. There they made us kindly welcome, did all they could for him, and transhipped the hammock into a pair-horse dray, which went quicker and was easier. We got on as fast as we could every step of the way, and by midnight that poor fellow was tucked into a clean bed in the hospital at Christchurch, with both his legs neatly cut off just above the knee, for there wasn't a minute to lose."

I was almost afraid to inquire how the sufferer fared, for Ned's eyes were fairly swimming with unshed tears; but he smiled brightly, and said, "The ladies and gentlemen in the town, they set up a subscribetion, and bought the poor chap a first-rate pair o' wooden legs, and he could even manage to ride about after a bit; and instead o' wandering about looking for country, or gold, or what not, he settled down as a carrier, and throve and did well. And I was thinking, ma'am, as how I'd like to return thanks for that poor fellow's wonderful recovery, for I've never had a chance of going to Church since, and its nigh upon two years ago that it happened."

"So you shall, Ned: so you shall!" we said with one voice. And so at our first Church gathering at our dear little antipodean home, F— —, who acted as our minister, paused in the beautiful Thanksgiving Service, after he had read solemnly and slowly the simple words, "Especially for Thy late mercies vouchsafed to — —," and Ned Palmer chimed in with an "Amen,"—misplaced, indeed, but none the less hearty, and delightful to hear.

Chapter X: Swaggers.

Dr. Johnson did not know the somewhat vulgar word which heads this paper. At least he did not know it as a noun, but gives "swagger: v.n., to bluster, bully, brag;" but the Slang Dictionary admits it as a word, springing indeed from the thieves' vocabulary: "one who carries a swag." Neither of these books however give the least idea of the true meaning of the expression, which is as fully recognised as an honest word in both Australia and New Zealand as any other combination of letters in the English language. A swagger is the very antithesis then of a swaggerer, for, whereas, the one is full of pretension and abounds in unjust claims on our notice, the swagger is humility and civility itself. He knows, poor weary tramp, that on the favourable impression he makes upon the "boss," depends his night's lodging and food, as well as a job of work in the future. We will leave then the ideal swaggerer to some other biographer who may draw glowing word-pictures of him in all his jay's splendour, and we will confine ourselves to describing the real swagger, clad in flannel shirt, moleskin trowsers, and what were once thick boots, but might now be used as sieves.

Nothing astonished me so much in my New Zealand Station Life as these visitors. Even Sir Roger de Coverley himself would have looked with distrust upon most of our swagger-guests, and yet I never heard of an instance in our part of the country where the unhesitating, ungrudging hospitality extended by the rich squatters to their poorer compatriots was ever abused. I say "in our part," because unfortunately, wherever gold is discovered, either in quartz or riverbed, the good old primitive customs and ways die out of themselves in a few weeks, and each mammon-seeker looks with distrust on a stranger. Only fifty or sixty miles from us, as the crow might fly across the snowy range, where an immense Bush clothes the banks of the Hokitika river right down to its sand-filled mouth on the West Coast, the great gold diggings broke out seven or eight years ago, and changed the face of society in that district in a few days. There a swagger meant a man who might rob or murder you in your sleep after you had fed and lodged him; or—under the most favourable circumstances supposing him to be a "milder mannered man," — a "fossicker," who would not hesitate to "jump your claim," or hang about when you are prospecting, to watch how much of the colour you found, and then go off stealthily to return next day at the head of a "rush" of a thousand diggers.

Even before the famous Maungatapu murders in 1866, swaggers were looked upon with distrust on the West Coast, and after that date hardly any one travelled in those parts without carrying a small revolver in his breast-pocket. Nothing is more tantalising than an allusion to a circumstance which is not well-known; and as I feel certain that very few of my readers have ever heard of what may be called the first great crime committed in the Middle Island, a brief account of that terrible tragedy may not be out of place. Gold of course was at the bottom of it, but the canvas-bags full of the glittering flakes were red with blood by the time they reached the bank at Nelson. The diggings on the West Coast were only two years old at that date, and although it was not uncommon for prospecting parties cutting their way, axe in hand, through the thick bush, to come upon skeletons of men in lonely places, still it might be taken for granted that these were the remains of early explorers or travellers who had got lost and starved to death within the green tangled walls of this impenetrable forest. The scenery of that part of the Middle Island is far more beautiful than in the agricultural or pastoral districts. Giant Alps clothed half up their steep sides with evergreen pines, — whose dark forms end abruptly where snow and ice begin, — stand out against a pure sky of more than Italian blue, and only when a cleared saddle is reached can the traveller look down over the wooded hills and vallies rolling away inland before him, or turn his eyes sea-ward to the bold coast with its many rivers, whose wide mouths foam right out to where the great Pacific waves are heaving under the bright winter sun.

Such, and yet still more fair must have been the prospect on which Burgess, Kelly, Levy, and Sullivan's eyes rested one June morning in the mid-winter of 1866. They were, one and all, originally London thieves, and had been transported years before to the early penal settlements of Australia. From thence they had managed, by fair means and foul, to work their way to other places, and had latterly been living in the Middle Island, earning what they could by horse-breaking and divers odd jobs. But your true convict hates work with a curiously deadly hatred, and these four men agreed to go and look round them at the new West Coast diggings. They found, however, that there, as elsewhere, it would be necessary to work hard, so in disgust at seeing the nuggets and dust which rewarded the toil of more industrious men, they left Hokitika and reached Nelson on their way to Picton, the chief town of the adjoining province of Marlborough. Most of the gold found its way under a strongly armed escort to the banks in both these towns, but it was well-known that fortunate diggers occasionally travelled together, unarmed, and laden with "dust." So safe had been the roads hitherto, that the commonest precautions were not taken, nor the least secrecy observed about travellers' movements.

It was therefore no mystery that four unarmed diggers, carrying a considerable number of ounces of gold-dust with them, were going to start from the Canvas-town diggings for Nelson on a certain day, and the men I have mentioned set out to meet them. One part of their long journey led them over the Maungatapu range by a saddle, which in its lowest part is 2,700 feet above the sea-level. The night before the murder, the victims and their assassins camped out with only ten miles between them. So lonely and deserted was the rough mountain track, that the appearance of a poor old man named Battle alarmed Burgess and his gang dreadfully, and they immediately murdered him, in order that he should not report having passed them on the road. Between the commission of this act of precaution and the arrival of the little band of travellers, no one else was seen. Burgess appears to have shown some of the qualities of a good general; for he selected a spot where the only path wound along a steep side-cutting, less than six feet wide, with an unbroken forest on the upper, and a mass of tangled bush on the lower side. As the doomed men approached the murderers sprang out, and each thrusting a revolver close to their faces, called on them "to hold up their hands." This is an old bushranger challenge, and is meant to ensure perfect quiescence on the part of the victim. The travellers mechanically complied, and in this way were instantly separated, led to different spots, and ruthlessly shot dead.

It was all over in a moment: Burgess and his men flung the bodies down among the tangled bush, and returned to Nelson rejoicing exceedingly over the simple and easy means by which they had possessed themselves of several hundred pounds. Of course they calculated on the usual supine indifference to other people's affairs, which prevails in busy gold-seeking communities; but in this instance the public seemed to be suddenly seized by a violent and inconvenient curiosity to find out what had become of the four men who were known to have started from Canvas-town two or three days before. No one ever dreamed of a murder having been committed, not even when another "swagger" reached Nelson and stated that he had followed the diggers on the road, only a mile or so behind, had suddenly lost sight of them at the spot I have mentioned, and had never been able to overtake them. Instead of leaving the now excited little town, or keeping quiet, Burgess, Kelly, Levy, and Sullivan, may truly be said to have become "swaggerers;" for they loitered about the place, ostentatiously displaying their bags of gold dust. Unsuspicious as the Nelson people were, they acted upon a sort of instinct,—that instinct within us which answers so mysteriously to the cry of blood from the earth,—and arrested these four men. Still, the matter might have ended there for lack of a clue, if one of the party, Sullivan, had not suddenly turned informer, and led the horrified town's-people to the jungle which concealed the bodies. Here my dreadful story may end; for we need not follow the course of the trial, which resulted in the complete conviction of the three other men. I have only dwelt on so horrible a theme in order to make my readers understand how natural it was that I should feel nervous, when it became apparent to my understanding that the custom of the country demanded that you should ask no questions, but simply tell any travellers who claimed your hospitality where they were to sleep, and send them in large supplies of mutton, flour, and tea.

On one occasion it chanced that F——, our stalwart cadet Mr. A——, and the man who did odd jobs about the place, were all on the point of setting out upon some expedition, when a party of four swaggers made their appearance just at sundown. No true swagger ever appears earlier, lest he might be politely requested to "move on" to the next station; whereas if he times his arrival exactly when "the shades of night are falling fast," no boss could be hard-hearted enough to point to mist-covered hills and valleys, which are a net-work of deep creeks and swamps, and desire the wayfarer to go on further. Once, and only once, did I know of such a thing being done; but I will not say more about that unfortunate at this moment, for I want to claim the pity of all my lady readers for the very unprotected position I am trying to depict. F—— could not understand my nervousness, and did not reassure me by saying, as he mounted his horse, "I've told them to sleep in the stable. I am pretty sure they are run-away sailors, they seem so footsore. Good-bye! don't expect me until you see me!"

Now I was a very new chum in those days, and had just heard of the Maungatapu murders. These guests of mine looked most disreputable, and were all powerful young men. I do not believe there was a single lock or bolt or bar on any door in the whole of the little wooden house: the large plate-chest stood outside in the verandah, and my dressing-case could have been carried off through the ever-open bedroom window by an enterprising thief of ten years old. As for my two maids,—the only human beings within reach,—they were as perfectly useless on any emergency as if they had been wax dolls. One of them had the habit of fainting if anything happened, and the other used to tend her until she revived, when they both sat still and shrieked. Their nerves had once been tested by a carpenter, who was employed about the house, and cut his hand badly; on another occasion by the kitchen chimney which took fire; and that was the way they behaved each time. So it was useless to look upon their presence as any safeguard; indeed one of them speedily detected a fancied likeness to Burgess in one of the poor swaggers, and shrieked every time she saw him.

We were indeed three "lone, 'lorn women," all through that weary night. I could not close my eyes; but laid awake listening to the weka's shrill call, or the melancholy cry of the bitterns down in the swamp. With the morning light came hope and courage; and I must say I felt ashamed of my suspicions when my cook came to announce that the "swaggers was just agoin' off, and wishful to say good-bye. They've been and washed up the tin plates and pannikins and spoons as clean as clean can be; and the one I thought favoured Burgess so much, mum, he's been and draw'd water from the well, all that we shall want to-day; and they're very civil, well-spoken chaps, if you please, mum!" F— — was right in his surmise, I fancy; for there were plenty of tattooed pictures of anchors and ships on the brawny bare arms of my departing guests. They seemed much disappointed to find there was no work to be had on our station; but departed, with many thanks and blessings, "over the hills and far away."

Latterly, with increasing civilization and corresponding social economy, there have been many attempts made by new-fangled managers of runs, more than by the run-holders themselves, to induce these swaggers to work for their tucker,—to use pure colonial phraseology. Several devices have been tried, such as taking away their swags (i.e., their red blankets rolled tightly into a sort of pack, which they carry on their backs, and derive their name from), and locking them up until they had chopped a small quantity of wood, or performed some other trifling domestic duty. But the swagger will be led, though not driven, and what he often did of his own accord for the sake of a nod or a smile of thanks from my pretty maid-servants, he would not do for the hardest words which ever came out of a boss's mouth. There are also strict rules of honesty observed among these men, and if one swagger were to purloin the smallest article from a station which had fed and sheltered him, every other swagger in all the country side would immediately become an amateur detective to make the thief give up his spoil. A pair of old boots was once missing from a neighbouring station, and suspicion fell upon a swagger. Justice was perhaps somewhat tardy in this instance, as it rested entirely in the hands of every tramp who passed that way; but at the end of some months the boots were found at home, and the innocence of the swaggers, individually and collectively, triumphantly established.

The only instance of harshness to a swagger which came under my notice during three years residence in New Zealand, is the one I have alluded to above, and contains so much dramatic interest in its details, that it may not be out of place here.

Although I have naturally dwelt in these papers more upon our bright sunny weather, our clear, bracing winter days, and our balmy spring and autumn evenings, let no intending traveller think that he will not meet with bad weather at the Antipodes! I can only repeat what I have said with pen and voice a hundred times before. New Zealand possesses a very capricious and disagreeable climate: disagreeable from its constant high winds: but it is perhaps the most singularly and remarkably healthy place in the world. This must surely arise from the very gales which I found so trying to my temper, for damp is a word without meaning; as for mildew or miasma, the generation who are growing up there will not know the meaning of the words; and in spite of a warm, bright day often turning at five minutes warning into a snowy or wet afternoon, colds and coughs are almost unknown. People who go out there with delicate lungs recover in the most surprising manner; surprising, because one expects the sudden changes of temperature, the unavoidable exposure to rain and even snow, to kill instead of curing invalids. But the practice is very unlike the theory in this case, and people thrive where they ought to die.

During my first winter in Canterbury we had only one week of really bad weather, but I felt at that time as if I had never realized before what bad weather meant. A true "sou'-wester" was blowing from the first to the second Monday in that July, without one moment's lull. The bitter, furious blast swept down the mountain gorges, driving sheets of blinding rain in a dense wall before it. Now and then the rain turned into large snow-flakes, or the wind rose into such a hurricane that the falling water appeared to be flashing over the drenched earth without actually touching it. Indoors we could hardly hear ourselves speak for the noise of the wind and rain against the shingle roof. It became a service of danger, almost resembling a forlorn hope, to go out and drag in logs of wet wood, or draw water from the well,—for, alas, there were no convenient taps or snug coal-holes in our newly-erected little wooden house. We husbanded every scrap of mutton, in very different fashion to our usual reckless consumption, the consumption of a household which has no butcher's bill to pay; for we knew not when the shepherd might be able to fight his way through the storm, with half a sheep packed before him, on sturdy little "Judy's" back. The creeks rose and poured over their banks in angry yellow floods. Every morning casualties in the poultry yard had to be reported, and that week cost me almost as many fowls and ducks as my great christening party did. The first thing every morning when I opened my eyes I used to jump up and look out of the different windows with eager curiosity, to see if there were any signs of a break in the weather, for I was quite unaccustomed to be pent up like a besieged prisoner for so many succeeding days. We did not boast of shutters in those regions, and even blinds were a luxury which were not wasted in the little hall. Consequently, when my unsatisfactory wanderings about the silent house—for no one else was up—led me that dreadful stormy morning into the narrow passage called the back-hall, I easily saw through its glass-door what seemed to me one of the most pathetic sights my eyes had ever rested upon.

Just outside the verandah, which is the invariable addition to New Zealand houses, stood, bareheaded, a tall, gaunt figure, whose rain-sodden garments clung closely to its tottering limbs. A more dismal morning could not well be imagined: the early dawn struggling to make itself apparent through a downpour of sleet and rain, the howling wind (which one could almost see as it drove the vapour wall before it), and the profound solitude and silence of all except the raging storm.

At first I thought I must be dreaming, so silent and hopeless stood that weird figure. My next impulse, without staying to consider my dishevelled hair and loose wrapper, was to open the door and beckon the poor man within the shelter of the verandah. When once I had got him there I did not exactly know what to do with my guest, for neither fire nor food could be procured quite so early. He crouched like a stray dog down on the dripping mat outside the door, and murmured some unintelligible words. In this dilemma I hastened to wake up poor F——, who found it difficult to understand why I wanted him to get up at daylight during a "sou'-wester." But I entreated him to go to the hall door, whilst I flew off to get my lazy maids out of their warm beds. With all their faults, they did not need much rousing on that occasion. I suppose I used very forcible words to convey the misery of the object standing outside, for I know that Mary was in floods of tears, and had fastened her gown on over her night-gear, whilst I was still speaking; and the cook had tumbled out of bed, and was kneeling before the kitchen fire with her eyes shut, kindling a blaze, apparently, in her sleep.

As soon as things were in this forward state, I returned to the verandah, and found our swagger guest drawing a very long breath after a good nip of pure whisky which F— — had promptly administered to him. "I'm fair clemmed wi' cold and wet," the swagger said, still bundled up in his comparatively sheltered corner. "I've been out on the hills the whole night, and I am deadbeat. Might I stop here for a bit?" He asked this very doubtfully, for it is quite against swagger etiquette to demand shelter in the morning. For all answer he was taken by the shoulder, and helped up. I never shall forget the poor tramp's deprecating face, as he looked back at me, whilst he was being led through the pretty little dining-room, with its bright carpet, on which his clay-clogged boots and dripping garments left a muddy, as well as a watery track. "All right," I said, with colonial brevity; and so we escorted our strange guest through the house into the kitchen, where the ever-ready kettle and gridiron were busy preparing tea and chops over a blazing fire. Of course the maids screamed when they saw us, and I do not wonder at their doing so, for neither F— — nor I looked very respectable, with huddled on dressing-gowns and towzled hair; whilst our foot-sore, drenched guest subsided into a chair by the door, covered his wretched pinched face with two bony hands, and burst into tears. I certainly never expected to see a swagger cry, and F— — declared the sight was quite as new to him as to me. However, the poor man's tears and helplessness gave fresh energy to my maids' treacherous nerves, and they even suggested dry clothes. Our good-natured cadet, who at this moment appeared on the scene, was only too happy to find some outlet for his superfluous benevolence, and hastened off, to return in a moment or two with an old flannel shirt, dry and whole, in spite of its faded stripes, a pair of moleskin trousers, and a huge pair of canvas cricketing shoes. It was no time for ceremony, so we women retreated for a few minutes into the store-room, whilst F— — and Mr. A— — made the swagger's toilette, getting so interested in their task as even to part his dripping hair out of his eyes. He had no swag, poor fellow, having lost his roll of red blankets in one of the treacherous bog-holes across the range.

That man was exactly like a lost, starving dog. He ate an enormous breakfast, curled himself upon some empty flour-sacks in a dry corner of the kitchen, and slept till dinner time; then another sleep until the supper hour, and so on, the round of he clock. All this time he never spoke, though we were dying to hear how he had come into such a plight. The "sou'-wester" still raged furiously out of doors without a moment's cessation, and we were obliged to have recourse to the tins of meat kept in the store-room for such an emergency. The shepherd told us afterwards he had ventured out to look for some wethers, his own supply being exhausted, but the whole mob had hidden themselves so cleverly that neither man nor dog could discover their place of shelter. On the Monday night, exactly a week after the outbreak of bad weather; the skies showed signs of having exhausted themselves, and nature began to wear a sulky air, as if her temper were but slowly recovering herself. The learned in such matters, however, took a cheerful view of affairs, and declared the worst to be over,—"for this bout,"—as they cautiously added.

Whether it was the three days of rest, warmth, and good food which unlocked the swagger's heart, or not, I do not pretend to decide; but that evening, over a pipe in the kitchen, he confided to Mr. A— —that he had been working his way down to the sea-coast from a station where he had been employed, very far back in the hill ranges. The "sou'-wester" had overtaken him about twenty miles from us, but only five from another station, where he had applied towards the evening for shelter, being even then drenched with rain, and worn out by struggling through such a tremendous storm. There, for some reason which I confess did not seem very clear, he had been refused the unvarying hospitality extended in New Zealand to all travellers, rich or poor, squatter or swagger, and had been directed to take a short cut across the hills to our station, which he was assured could easily be reached in an hour or two more. The track, a difficult one enough to strike in summer weather, became, indeed, impossible to discover amid rushing torrents and driving wind and rain; besides which, as the poor fellow repeated more than once during his story, "I was fair done up when I set out, for I'd been travelling all day." Mr. A—— told us what the man had been saying, before we all went to bed, adding, "He seems an odd, surly kind of creature, for although he declares he is going away the first thing to-morrow, if the rain be over, I noticed he never said a word approaching to thanks."

The rain was indeed over next morning, and a flood of brilliant sunshine awoke me "bright and early," as the country people say. It seemed impossible to stop in bed, so I jumped up, thrust my feet into slippers, and my arms into a warm dressing-gown, and sallied forth, opening window after window, so as to let the sunshine into rooms which not even a week's steady down-pour could render damp. What a morning it was, and for mid-winter too! No haze, or fog, or vapour on all the green hills, whose well-washed sides were glistening in a bright glow of sunlight. For the first time, too, since the bad weather had set in, was to be heard the incessant bleat which is music to the ears of a New Zealand sheep-farmer. White, moving, calling patches on the hillsides told that the sheep were returning to their favourite pastures, and a mob of horses could be descried quietly feeding on the sunny flat.

But I had no eyes for beauties of mountain or sky. I could do nothing but gaze on the strange figure of the silent swagger, who knelt yes, positively knelt, on the still wet and shining shingle which formed an apology for a gravel path up to the back-door of the little wooden homestead. His appearance was very different to what it had been three days before. Now his clothes were dry and clean and mended,—my Irish maids doing; bless their warm hearts! He had cobbled up his boots himself, and his felt hat, which had quite recovered from its drenching, lay at his side. The perfect rest and warmth and good food had filled up his hollow cheeks, but still his countenance was a curious one; and never, until my dying day, can I forget the rapture of entreaty on that man's upturned face. It brings the tears into my own eyes now to recollect its beseeching expression. I do not think I ever saw prayer before or since. He did not perceive me, for I had hidden behind a sheltering curtain, to listen to his strange, earnest petitions; so he could not know that anybody in the house was stirring, for he knelt at the back, and all my fussings had taken place in the front, and he could not, therefore, have been doing anything for effect.

There, exactly where he had crouched a wretched, way-worn tramp in pouring rain, he knelt now with the flood of sunshine streaming down on his uplifted face, whilst he prayed for the welfare and happiness, individually and collectively, of every living creature within the house. Then he stood up and lifted his hat from the ground; but before he replaced it on his head, he turned, with a gesture which would have made the fortune of any orator,—a gesture of mingled love and farewell, and solemnly blessed the roof-tree which had sheltered him in his hour of need. I could not help being struck by the extraordinarily good language in which he expressed his fervent desires, and his whole bearing seemed quite different to that of the silent, half-starved man we had kept in the kitchen these last three days. I watched him turn and go, noiselessly closing the garden gate after him, and—shall I confess it?—my heart has always felt light whenever I think of that swagger's blessing. When we all met at breakfast I had to take his part, and tell of the scene I had witnessed; for everybody was inclined to blame him for having stolen away, scarcely without saying good-bye, or expressing a word of thanks for the kindness he had received. But I knew better.

From the sublime to the ridiculous we all know the step is but short, especially in the human mind; and to my tender mood succeeds the recollection of an absurd panic we once suffered from, about swaggers. Exaggerated stories had reached us, brought by timid fat men on horseback, with bulky pocket-books, who came to buy our wethers for the Hokitika market, of "sticking up" having broken out on the west land. I fear my expressions are often unintelligible to an English reader, but in this instance I will explain. "Sticking up" is merely a concise colonial rendering of "Your money or your life," and was originally employed by Australian bushrangers, those terrible freebooters whose ranks used to be always recruited from escaped convicts. Fortunately we had no community of that class, only a few prisoners kept in a little ricketty wooden house in Christchurch, from which an enterprising baby might easily have escaped. I dare say as we get more civilized out there, we shall build ourselves handsome prisons and penitentiaries; but in those early days a story was current of a certain jailor who let all his captives out on some festal occasion, using the tremendous threat, that whoever had not returned by eight o'clock should be "locked out!"

But to return to that particular winter evening. We had been telling each other stories which we had heard or read of bushranging exploits, until we were all as nervous as possible. Ghosts, or even burglar stories, are nothing to the horror of a true bushranger story, and F—— had made himself particularly ghastly and disagreeable by giving a minute account of an adventure which had been told to him by one of the survivors.

We listened, with the wind howling outside, to F——'s horrid second-hand story, of how one fine day up country, eight or ten men,—station hands,—were "stuck up" by one solitary bushranger, armed to the teeth. He tied them up one by one, and seated them all on a bench in the sun, and deliberately fired at and wounded the youngest of the party; then, seized with compunction, he unbound one of the captives, and stood over him, revolver in hand, whilst he saddled and mounted a horse, to go for a doctor to set the poor boy's broken leg. Before the messenger had gone "a league, a league, but barely twa',"—the freebooter recollected that he might bring somebody else back with him besides the doctor, and flinging himself across his horse, rode after the affrighted man, and coolly shot him dead. I really don't know how the story ended: I believe everybody perished; but at this juncture I declared it to be impossible to sit up any longer to listen to such tragedies, and went to bed.

Exactly at midnight,—the proper hour for ghosts; burglars, and bushrangers, and such "small deer" to be about, everybody was awakened simultaneously by a loud irregular knocking, which sounded with hollow reverberations all through the wooden house. "Bushrangers!" we all thought, every one of us; for although burglars may not usually knock at hall-doors in England, it is by no means uncommon for their bolder brethren to do so at the other end of the world. It is such a comfort to me now, looking back on that scene to remember that our stalwart cadet was as frightened as anybody. He stood six feet one in his stockings, and was a match for any two in the country side, and yet, I am happy to think, he was as bad as any one. As for me, to say that my heart became like water and my knees like soft wax, is to express in mild words my state of abject terror. There was no need to inquire what the maids thought, for smothered shrieks, louder and louder as each peal of knocks vibrated through the little house, proclaimed sufficiently their sentiments on the subject.

Dear me, how ridiculous it all must have been! In one corner of the ceiling of our bedroom was a little trap-door which opened into an attic adjoining that where the big cadet slept. Now whilst F—— was hurriedly taking down his double-barrelled gun from its bracket just below this aperture, and I held the candlestick with so shaky a hand that the extinguisher clattered like a castanet, this door was slowly lifted up, and a large white face, with dishevelled stubbly hair and wide-open blue eyes, looked down through the cobwebs, saying in a husky whisper, "Could you let me have a rifle, or any thing?" This was our gallant cadet, who had no idea of presenting himself at a disadvantage before the foe. I had desperately seized a revolver, but F—— declared that if I persisted in carrying it I certainly should go first, as he did not wish to be shot in the back.

We held a hurried council of war,—Mr. A—— assisting through the trap door, and the maids breathing suggestions through the partition-planks,—but the difficulty consisted in determining at which door the knocking was going on. Some said one, and some another (for there were many modes of egress from the tiny dwelling); but at last F—— cried decidedly, "We must try them all in succession," and shouldering his gun, with the revolver sticking in the girdle of his dressing-gown, sallied valiantly forth. I don't know what became of Mr. A——: I believe he took up a position with the rifle pointing downwards; the maids retreated beneath their blankets, and I (too frightened to stay behind) followed closely, armed with an Indian boar-spear. F— — flung the hall door wide open, and called out, "Who's there?" but no one answered. The silence was intense, and so was the cold; therefore we returned speedily indoors to consult. "It must be at the back door," I urged; adding, "that is the short cut down the valley, where bushrangers would be most likely to come." "Bushrangers, you silly child!" laughed F— —. "It's most likely a belated swagger, or else somebody who is playing us a trick." However as he spoke a succession of fierce and loud knocks resounded through the whole house. "It must be at the kitchen door," F— — said. "Come along, and stand well behind me when I open the door."

But we never opened the door; for on our way through the kitchen, with its high-pitched and unceiled roof,—a very cavern for echoes,—we discovered the source of the noise, and of our fright. Within a large wooden packing-case lay a poor little lamb, and its dying throes had wakened us all up, as it kicked expiring kicks violently against the side of the box. It was my doing bringing it indoors, for I never could find it in my heart to leave a lamb out on the hills if we came across a dead ewe with her baby bleating desolately and running round her body. F— — always said, "You cannot rear a merino lamb indoors; the poor little thing will only die all the same in a day or two;" and then I am sorry to say he added in an unfeeling manner, "They are not worth much now," as if that could make any difference! I had brought this, as I had brought scores of others, home in my arms from a long distance off; fed it out of a baby's bottle, rubbed it dry, and put it to sleep in a warm bed of hay at the bottom of this very box. They had all died quietly, after a day or two, in spite of my devotion and nursing, but this little foundling kicked herself out of the world with as much noise as would have sufficed to summon a garrison to surrender. It is all very well to laugh at it now, but we were, five valiant souls in all, as thoroughly frightened at the time as we could well be.

The only real harm a swagger did me was to carry off one of my best maidservants as his wife, but as he had 300 pounds in the bank at Christchurch, and was only travelling about looking for work, and they have lived in great peace and prosperity ever since, I suppose I ought not to complain. This swagger was employed in deepening our well, and Mary was always going to see how he was getting on, so he used to make love to her, looking up from the bottom of a deep shaft, and shouting compliments to her from a depth of sixty feet. What really won her Irish heart, though, was his calmly putting a rival, a shepherd, into a water-butt. She could not resist that, so they were married, and are doing well.

Let no one despise swaggers. They are merely travelling workmen, and would pay for their lodging if it was the custom to do so. I am told that even now they are fast becoming things of the past; for one could not "swagger" by railroad, and most of our beautiful happy vallies will soon have a line of rails laid down throughout its green and peaceful length.

Chapter XI: Changing servants.

To the eyes of an English housewife the title of this chapter must appear a very bad joke indeed, and the amusement what the immortal Mrs. Poyser would call "a poor tale." Far be it from me to make light of the misery of a tolerably good servant coming to you after three months' service, just as you were beginning to feel settled and comfortable, and announcing with a smile that she was going to be married; or, with a flood of tears, that she found it "lonesome." Either of these two contingencies was pretty sure to arise at least four times a year on a station.

At first I determined to do all I could to make their new home so attractive to my two handmaidens that they would not wish to leave it directly. In one of Wilkie Collins' books an upholsterer is represented as saying that if you want to domesticate a woman, you should surround her with bird's-eye maple and chintz. That must have been exactly my idea, for the two rooms which I prepared for my maidservants were small, indeed, yet exquisitely pretty. Of course I should not have been so foolish as to buy any of the unnecessary and dainty fittings with which they were decorated, but as all the furniture and belongings of an English house, a good deal larger than our station home, had been taken out to it, there were sundry toilet tables, etc., whose destination would have been a loft over the stable, if I had not used them for my maids.

I had seen and chosen two very respectable young women in Christchurch, one as a cook, and the other as a housemaid. The cook, Euphemia by name, was a tall, fat, flabby woman, with a pasty complexion, but a nice expression of face, and better manners than usual. She turned out to be very good natured, perfectly ignorant though willing to learn, and was much admired by the neighbouring cockatoos, or small farmers. Lois the housemaid, was the smallest and skimpiest and most angular girl I ever beheld. At first I regarded her with deep compassion, imagining that she was about fifteen years of age, and had been cruelly ill-treated and starved. How she divined what was passing in my mind I cannot tell, but during our first interview she suddenly fired up, and informed me that she was twenty-two years old, that she was the seventh child of a seventh child, and therefore absolutely certain to achieve some wonderful piece of good luck; and furthermore, that she had been much admired in her own part of the country, and was universally allowed to be "the flower of the province." This statement, delivered with great volubility and defiant jerkiness of manner, rather took my breath away; but it was a case of "Hobson's choice" just then about servants, and as I was assured she was a respectable girl, I closed with her terms (25 pounds a year and all found) on the spot. The fat pale cook was to get 35 pounds. Now-a-days I hear that wages are somewhat lower, but the sums I have named were the average figures of six or seven years ago, especially "up-country."

Here I feel impelled to repeat the substance of what I have stated elsewhere,—that these rough, queer servants were, as a general rule, perfectly honest, and of irreproachable morals, besides working, in their own curious fashion, desperately hard. Our family was an exceptionally small one, and the "place" was considered "light, you bet," but even then it seemed to me as if both my domestics worked very hard. In the first place there was the washing; two days severe work, under difficulties which they thought nothing of. All the clothes had to be taken to a boiler fixed in the side of a hill, for the convenience of the creek, and washed and rinsed under a blazing sun (for of course it never was attempted on a wet day) and amid clouds of sand-flies. Not until evening was this really hard day's work over, and the various garments fluttering in the breeze up a valley behind the house. The chances were strongly in favour of a tremendous nor'-wester coming down this said valley during the night, and in that case there would not be a sign next morning of any of the clothes. Heavy things, such as sheets or table cloths, might be safely looked for under lee of the nearest gorse hedge, but it would be impossible even to guess where the lighter and more diaphanous articles had been whisked to. A week afterwards the shepherds used to bring in stray cuffs and collars, and upon one occasion "Judy," the calf, was discovered in a paddock hard by, breakfasting off my best pocket handkerchiefs with an excellent appetite. Of course everything was dirty, and needed to be washed over again. We had a mangle, which greatly simplified matters on the second day, but it used not to be uncommon on back-country stations to get up the fine things with a flat stone, heated in the wood ashes, for an iron. After the washing operations had been brought to a more or less successful ending, there came the yeast making and the baking, followed by the brewing of sugar beer, preserves had to be made, bacon cured, all sorts of things to be done, besides the daily duties of scrubbing and cleaning, and cooking at all hours for stray visitors or "swaggers."

But I am overcome with contrition at perceiving into what a digression I have wandered; having strayed from my maids' rooms to their duties. They arrived as usual on a dray late in the evening, tired and wearied enough, poor souls. In those early days I had not yet plucked up courage to try my hand in the kitchen, and our meals had been left to the charge of F— —, who, whatever he may be in other relations of life, is a vile cook; and our good-natured cadet Mr. U— —, who was exceedingly willing, but profoundly ignorant of the elements of cookery. For fear of being tempted into another digression, I will briefly state that during that week I lived in a chronic state of hunger and heartburn, and sought forgetfulness from repeated attacks of indigestion, by decorating my servants' rooms. They opened into each other, and it would have been hard to find two prettier little nests. Each had its shining brass bedstead with chintz hangings, its muslin-draped toilette table, and its daintily curtained window, besides a pretty carpet. I can remember now the sort of dazed look with which Euphemia regarded a room such as she had never seen; whilst Lois considered it to be an instalment of her good luck, and proceeded to contemplate her sharp and elfish countenance in her looking-glass, pronouncing it as her opinion that she wanted more colour. That she certainly did, and she might have added, more flesh and youthfulness, while she was about it. However, they were greatly delighted, and Euphemia who was of a grateful and affectionate disposition, actually thanked me, for having with my own hands arranged such pretty rooms for them.

This was a very good beginning. They were both hard-working, civil girls, and got on very well together, leaving me plenty of leisure to attend to the quantities of necessary arrangements which have to be made when you are settling yourself for good, fifty miles from a shop, and on a spot where no other human being has ever lived before. F— — congratulated myself in private on my exceptional good luck, and attributed it partly to my having followed the Upholsterer's advice in that book of Mr. Wilkie Collins. But as it turned out, F— — was dwelling in a fool's paradise. In vain had it been pointed out to me that a certain stalwart north countryman, whose shyness could only be equalled by his appetite, had been a most regular attendant for some weeks past at our Sunday evening services, accepting the offer of tea in the kitchen, afterwards, with great alacrity. I scouted these insinuations, appealing to the general sense of the public as to whether Moffatt had ever been known to refuse a meal anywhere, or under any circumstances, and declaring that, if he was "courting," it was being done in solemn silence, for never a sound filtered through the thin wooden planks between the kitchen and the dining room, except the clatter of a vigorously plied knife and fork, for Moffatt's teas always included a shoulder of mutton.

But I was wrong and others were right. Early in October, our second spring month, I chanced to get up betimes one delicious, calm morning, a morning when it seemed a new and exquisite pleasure to open each window in succession, and fill one's lungs with a deep, deep breath of that heavenly atmosphere, at once so fresh and so pure.

Quiet as the little homestead lay, nestled among the hills, there were too many morning noises stirring among the animals for any one to feel lonely or dull, I should have thought. From a distance came a regular, monotonous, lowing sound. That was "Hetty," the pretty little yellow Alderney, announcing from the swamps that she and her two female friends were quite ready to be milked. Their calves answered them dutifully from the English grass paddock, and between the two I could see Mr. U— —'s tall figure stalking down the flat with his cattle dog at his heels, and hear his merry whistle shrilling through the silent air. Then all the ducks and fowls about the place were inquiring, in noisy cackle, how long it would be before breakfast was ready, whilst "Helen's" whinneying made me turn my head to see her, with a mob of horses at her heels, coming over the nearest ridge on the chance of a stray carrot or two going begging. All the chained-up dogs were pulling at the staples of their fastenings, and entreating by short, joyous barks, to be allowed just one good frisk and roll in the sparkling dewy grass around. But even I, universal spoiler of animals that I am, was obliged to harden my heart against their noisy appeals; for quite close to the stable, on the nearest hill-side, an immense mob of sheep and young lambs were feeding. That steep incline had been burnt six weeks before, and was now as green as the clover field at its base, affording a delicious pasturage to these nursing mothers and their frisky infants. I think I see and hear it all now. The moving white patches on the hill-side, the incessant calling and answering, the racing and chasing among the curly little merino lambs, and above all the fair earth the clear vault of an almost cloudless sky bent itself in a deep blue dome. Just over the eastern hills the first long lances of the sun lay in bright shafts of silver sheen on the dew-laden tussocks, and that peculiar morning fragrance rose up from the moist ground, which is as much the reward of the early riser as the early worm is of the bird.

Was it a morning for low spirits or sobs and sighs? Surely not; and yet as I turned the handle of the kitchen door those melancholy sounds struck my ear. I had intended to make my entrance with a propitiatory smile, suitable to such a glorious morning, proceed to pay my damsels a graceful compliment on their somewhat unusual early rising, and wind up with a request for a cup of tea. But all these friendly purposes went out of my head when I beheld Euphemia seated on the rude wooden settle, with its chopped tussock mattrass, which had been covered with a bright cotton damask, and was now called respectfully, "the kitchen sofa." Her arm was round Lois's waist, and she had drawn that young lady's shock head of red curls down on her capacious bosom. Both were crying as if their hearts would break, and startled as I felt to see these floods of tears, it struck me how incongruous their attitude looked against the background of the large window through which all nature looked so smiling and sparkling. The kettle was singing on the fire, everything seemed bright and snug and comfortable indoors. "What in the world has happened?" I gasped, really frightened.

"Nothing, mem: its only them sheep," sobbed Euphemia, "calling like. They always makes me cry. Your tea 'll be ready directly, mem" (this last with a deep sigh.)

"Is it possible you are crying about that?" I inquired. "Yes, mem, yes," said Euphemia, in heart-broken accents, clasping Lois, who was positively howling, closer to her sympathetic heart. "Its terrible to hear 'em. They keeps calling and answering each other, and that makes us think of our home and friends." Now both these women had starved as factory "hands" all their lives, and I used to feel much more inclined to cry when they told me, all unconscious of the pathos, stories of their baby work and hardships. Certainly they had never seen a sheep until they came to New Zealand, and as they had particularly mentioned the silence which used to reign supreme at the manufactory during work hours, I could not trace the connection between a dingy, smoky, factory, and a bright spring morning in this delightful valley. "What nonsense!" I cried, half laughing and half angry. "You can't be in earnest. Why you must both be ill: let me give you each a good dose of medicine." I said this encouragingly, for there was nothing in the world Euphemia liked so much as good substantial physic, and the only thing I ever needed to keep locked up from her was the medicine drawer.

Euphemia seemed touched and grateful, and her face brightened up directly, but Lois looked up with her frightful little face more ugly than usual, as she said, spitefully, "Physic won't make them nasty sheep hold their tongues. I'm sure this isn't the place for me to find my luck, so I'd rather go, if you please, mem. I've prospected-up every one o' them gullies and never seen the colour yet, so it ain't any good my stopping."

This was quite a fresh light thrown upon the purpose of Lois's long lonely rambles. She used to be off and away, over the hills whenever she had finished her daily work, and I encouraged her rambles, thinking the fresh air and exercise must do her a world of good. Never had I guessed that the sordid little puss was turning over every stone in the creek in her search for the shining flakes.

"Why did you think you should find gold here?" I asked.

"Because they do say it lies in all these mountain streams," she answered sullenly; "and I'm always dreaming of nuggets. Not that a girl with my face and figure wants 'dust' to set her off, however. But if its all the same to you, mem, I'd rather leave when Euphemia does."

"Are you going, then?" I inquired, turning reproachfully to my pale-faced cook, who actually coloured a little as she answered, "Well, mem, you see Moffatt says he's got his window frames in now, and he'll glass them the very first chance, and I think it'll be more company for me on Saddler's Flat. So if you'll please to send me down in the dray, I should be obliged."

Here was a pretty upset, and I went about my poultry-feeding with a heavy heart. How was I to get fresh servants, and above all, what was I to do for cooking during the week they were away? These questions fortunately settled themselves in rather an unexpected manner. I heard of a very nice willing girl who was particularly anxious to come up as housemaid, to my part of the world, on condition that I should also engage as cook her sister, who was leaving a place on the opposite side of a range of high hills to the south. I shall only briefly say that all inquiries about these damsels proved satisfactory, and I could see Euphemia and Lois depart, with tolerable equanimity. The former wept, and begged for a box of Cockles' pills; but Lois tossed her elfish head, and gave me to understand that she had never been properly admired or appreciated whilst in my service.

Chapter XII: Culinary troubles.

I want to lodge a formal complaint against all cookery books. They are not the least use in the world, until you know how to cook! and then you can do without them. Somebody ought to write a cookery book which would tell an unhappy beginner whether the water in which she proposes to put her potatoes is to be hot or cold; how long such water is to boil; how she is to know whether the potatoes are done enough; how to dry them after they have boiled, and similar things, which make all the difference in the world.

To speak like Mr. Brooke for a moment. "Rice now: I have dabbled in that a good deal myself, and found it wouldn't do at all."

Of course in time, and after many failures, I did learn to boil a potato which would not disgrace me, and to bake bread, besides in time attaining to puddings and cakes, of which I don't mind confessing I was modestly proud. It used to be a study, I am told, to watch my face when a cake had turned out as it ought. Gratified vanity at the lavish encomiums bestowed on it, and horrified dismay at the rapidity with which a good sized cake disappeared down the throats of the company, warred together in the most artless fashion. The reflection would arise that it was almost a pity it should be eaten up so very fast; yet was it not a fine thing to be able to make such a cake! and oh, would the next be equally good?

One lesson I leaned in my New Zealand kitchen,—and that was not to be too hard on the point of breakages; for no one knows, unless from personal experience, how true was the Irish cook's apology for breaking a dish, when she said that it let go of her hand. I declare that I used, at last, to regard my plates and dishes, cups and saucers, yea, even the pudding basons, not as so much china and delf, but as troublesome imps, possessed with an insane desire to dash themselves madly on the kitchen floor upon the least provocation. Every woman knows what a slippery thing to hold is a baby in its tub. I am in a position to pronounce that wet plates and dishes are far more difficult to keep hold of. They have a way of leaping out of your fingers, which must be felt to be believed. After my first week in my kitchen I used to wonder, not at the breakages, but at anything remaining unbroken.

My maids had a very ingenious method of disposing of the fragments of their pottery misfortunes. At the back of the house an open patch of ground, thickly covered with an under-growth of native grass, and the usual large proportion of sheltering tussocks stretched away to the foot of the nearest hill. This was burned every second year or so, and when the fire had passed away the sight it revealed was certainly very curious. Beneath each tussock had lain concealed a small heap of broken china, which must have been placed there in the dead of the night. The delinquents had evidently been at the pains to perfect their work of destruction by reducing the china articles in question, to the smallest imaginable fragments, for fear of a protruding corner betraying the clever cache; and the contrast afforded to the blackened ground on which they lay, by the gay patches of tiny fragments huddled together, was droll indeed. That was the moment for recognising the remains of a favourite jug or plate, or even a beloved tea-cup. There they were all laid in neat little heaps, and the best of it was that the existing cook always declared loudly her astonishment at the base ingenuity of such conduct, although I could not fail to recognise many a plate or dish which had disappeared from the land of the living during her reign.

All housekeepers will sympathise with my feelings at seeing an amateur scullion, who had distinguished himself greatly in the Balaklava charge, but who appeared to have no idea that boiling water would scald his fingers,—drop the top plate of a pile which he had placed in a tub before him. In spite of my entreaties to be allowed to "wash-up" myself, he gallantly declared that he could do it beautifully, and that the great thing was to have the water very hot. In pursuance of this theory he poured the contents of a kettle of boiling water over his plates, plunged his hand in, and dropped the top plate, with a shriek of dismay, on those beneath it. Out of consideration for that well-meaning emigrant's feelings, I abstain from publishing the list of the killed and wounded, briefly stating that he might almost as well have fired a shot among my poor plates. A perfect fountain of water and chips and bits of china flew up into the air, and I really believe that hardly one plate remained uncracked. So much for one's friends. I must candidly state that although the servants broke a good deal, we destroyed twice as much amongst us during the week which must needs elapse between their departure and, the arrival of the new ones.

Shall I ever forget the guilty pallor which overspread the bronzed and bearded countenance of one of my guests, who particularly wished to dust the drawing-room ornaments, when on hearing a slight crash I came into the room and found him picking up the remains of a china shepherdess? Considering everything, I kept my temper remarkably well, merely observing that he had better go into the verandah and sit down with a book and his pipe, and send Joey in to help me. Joey was a little black monkey from Panama, who had to be provided with broken bits of delf or china in order that he might amuse himself by breaking them ingeniously into smaller fragments.

But the real object of this chapter was to relate some of my own private misfortunes in the cooking line. Once, when Alice S— — was staying with me and we had no servants, she and I undertook to bake a very infantine and unweaned pig. It was all properly arranged for us, and, making up a good fire, we proceeded to cook the little monster.

Hours passed by; all the rest of the dinner got itself properly cooked at the right time, but the pig presented exactly the same appearance at dewy eve as it had done in the early morn. We looked rather crest-fallen at its pale condition when one o'clock struck, but I said cheerfully, "Oh, I daresay it will be ready by supper!" But it was not: not a bit of it. Of course we searched in those delusive cookery books, but they only told us what sauces to serve with a roasted pig, or how to garnish it, entering minutely into a disquisition upon whether a lemon or an orange had better be stuck into its mouth. We wanted to know how to cook it, and why it would not get itself baked. About an hour before supper-time I grew desperate at the anticipation of the "chaff" Alice and I would certainly have to undergo if this detestable animal could not be produced in a sufficiently cooked state by evening. We took it out of the oven and contemplated it with silence and dismay. Fair as ever did that pig appear, and as if it had no present intention of being cooked at all. A sudden idea came into our heads at the same moment, but it was Alice who first whispered, "Let us cut off its head." "Yes," I cried; "I am sure that prevents its roasting or baking, or whatever it is." So we got out the big carving knife and cut off the piggy's head. Far be it from me to offer any solution of the theory why the head should have interfered with the baking process, but all I know is, that, like the old woman in the nursery song, everything began to go right, and we got our supper that night.

Has anybody ever reflected on how difficult it must be to get a chimney swept without ever a sweep or even a brush? Luckily our chimneys were short and wide, and we used a good deal of wood; so in three years the kitchen chimney only needed to be cleansed twice. The first time it was cleared of soot by the simple process of being set on fire, but as a light nor'-wester was blowing, the risk to the wooden roof became very great and could only be met by spreading wet blankets over the shingles. We had a very narrow escape of losing our little wooden house, and it was fortunate it happened just at the men's dinner hour when there was plenty of help close at hand. However great my satisfaction at feeling that at last my chimney had been thoroughly swept, there was evidently too much risk about the performance to admit of its being repeated, so about a year afterwards I asked an "old chum" what I was to do with my chimney. "Sweep it with a furze-bush, to be sure," she replied. I mentioned this primitive receipt at home, and the idea was carried out a day or two later by one man mounting on the roof of the house whilst another remained in the kitchen; the individual on the roof threw down a rope to the one below, who fastened a large furze-bush in the middle, they each held an end of this rope, and so pulled it up and down the chimney until the man below was as black as any veritable sweep, and had to betake himself, clothes and all, to a neighbouring creek. As for the kitchen, its state cannot be better described than in my Irish cook's words, who cried, "Did mortial man ever see sich a ridiklous mess? Arrah, why couldn't ye let it be thin?" But for all that she set bravely to work and got everything clean and nice once more, merely stipulating that the next time we were going to sweep chimney we should let her know beforehand, that she might go somewhere "right away."

I feel, however, that in all these reminiscences I am straying widely from the point which was before my mind when I began this chapter, and that is the delusiveness of a cookery book. No book which I have ever seen tells you, for instance, how to boil rice properly. They all insist that the grains must be white and dry and separate, but they omit to describe the process by which these results can be attained. They tell you what you are to do with your rice after it is boiled, but not how to boil it. The fact is, I suppose, that the people who write such books began so early to be cooks themselves, that they forget there ever was a time when such simple things were unknown to them.

Even when I had, after many failures, mastered the art of boiling rice, and also of making an excellent curry,—for which accomplishment I was indebted to the practical teaching of a neighbour,—there used still to be misfortunes in store for me. One of these caused me such a bitter disappointment that I have never quite forgotten it. This was the manner of it. We were without servants. My readers must not suppose that such was our chronic condition, but when you come to change your servants three or four times a year, and have to "do" for yourself each time during the week which must elapse before the arrival of new ones, there is an ample margin for every possible domestic misadventure. If any doubt me, let them try for themselves.

On this special occasion, which proved to be nearly the last, my mind was easy, for the simple reason that I was now independent of cookery books. I had puzzled out all the elementary parts of the science for myself, and had no misgivings on the subject of potatoes or even peas. So confident was I, and vain, that I volunteered to make a curry for breakfast. Such a savoury curry as it was, and it turned out to be all that the heart of a hungry man could desire; so did the rice: I really felt proud of that rice; each grain kept itself duly apart from its fellow, and was as soft and white and plump as possible. Everything went well, and I had plenty of assistants to carry in the substantial breakfast as fast as it was ready: the coffee, toast, all the other things had gone in; even the curry had been borne off amid many compliments, and now it only remained for me to dish up the rice.

Imagine the scene. The bright pretty kitchen, with its large window through which you could see the green hills around dotted with sheep; the creek chattering along just outside, whilst close to the back door loitered a crowd of fowls and ducks on the chance of fate sending them something extra to eat. Beneath the large window, and just in front of it, stood a large deal table, and it used to be my custom to transfer the contents of the saucepans to the dishes at that convenient place. Well, I emptied the rice into its dish, and gazed fondly at it for a moment: any cook might have been proud of that beautiful heap of snow-white grains. I had boiled a great quantity, more than necessary it seemed, for although the dish was piled up almost as high as it would hold, some rice yet remained in the saucepan.

Oh, that I had been content to leave it there! But no: with a certain spasmodic frugality which has often been my bane, I shook the saucepan vehemently, in order to dislodge some more of its contents into my already full dish. As I did so, my treacherous wrist, strained by the weight of the saucepan, gave way, and with the rapidity of a conjurer's trick I found the great black saucepan seated,—yes, that is the only word for it,—seated in the midst of my heap of rice, which was now covered by fine black powder from its sooty outside. All the rice was utterly and completely spoiled. I don't believe that five clean grains were left in the dish There was nothing for it but to leave it to get cold and then throw it all out for the fowls, who don't mind riz au noir it seems. Although I feel more than half ashamed to confess it, I am by no means sure I did not retire into the store-room and shed a tear over the fate of that rice. Everybody else laughed, but I was dreadfully mortified and vexed.

Chapter XIII: Amateur Servants.

I flattered myself on a certain occasion that I had made some very artful arrangements to provide the family with something to eat during the servants' absence. I had been lamenting the week of experiments in food which would be sure to ensue so soon as the dray should leave, in the hearing of a gallant young ex-dragoon, who had come out to New Zealand to try and see if one could gratify tastes, requiring, say a thousand a year to provide for, on an income of 120 pounds. He was just finding out that it was quite as difficult to manage this in the Southern as in the Northern Hemisphere, but his hearty cheery manner, and enormous stock of hope, kept him up for some time.

"I'll come and cook for you," he cried. "I can cook like a bird. But I can't wash up. No, no: it burns too much. If you can get somebody to wash up, I'll cook. And just look here: it would be very nice if we could have some music after dinner. You've got a piano, haven't you? That's right. Well, now, don't you ask that pretty Miss A— —, who has just come out from England, to come and stop with you, and then we could have some music?"

"Where did you learn to cook?" I inquired, suspiciously; for F— —had also assured me he could cook, and this had upset my confidence.

"On the west coast; to be sure! Ask Vere, and Williams and Taylor, and everybody, if they ever tasted such pies as I used to make them." My countenance must have still looked rather doubtful, because I well remember sundry verbal testimonials of capability being produced; and as I was still very ignorant of the rudiments of the science of cookery, I shrank from assuming the whole responsibility of the family meals. So the household was arranged in this way:—Captain George, head cook; Mr. U— —, scullery-maid; Miss A— —, housemaid; myself, lady-superintendent; Mr. Forsyth (a young naval officer), butler. On the principle of giving honour to whom honour is due, this gallant lieutenant deserves special mention for the way he cleaned glass. He did not pay much attention to his silver, but his glass would have passed muster at a club. The only drawback was the immense time he took over each glass, and the way he followed either Miss A— — or me all about the house, holding a tumbler in one hand, and a long, clean glass-cloth in the other, calling upon us to admire the polish of the crystal. To clean two tumblers would be a good day's work for him. From Monday to Saturday (when the dray returned), this state of things went on. Of course I had taken the precaution of having a good supply of bread made beforehand, besides cakes and biscuits, tarts and pies; everything to save trouble. But it was not of much use, for, alleging that they were working so hard, the young men, F— — at their head, though I was always telling him he was married and ought to know better, set to work and ate up everything immediately, as completely as if they had been locusts. And then, they were all so dreadfully wild and unmanageable! Mine was by far the hardest task of all, the keeping them in any sort of order. For instance, Captain George declared one day, that if there was one thing he did better than another, it was to make jam. Consequently a fatigue party was ordered out to gather strawberries, and, after more than half had been eaten on the way to the house, a stewpan was filled. I had to do most of the skimming, as Captain George wanted to practice a duet with Miss A— —. I may as well mention here that we never had any opportunity of seeing how the jam kept, because the smell pervaded the whole house to such an extent, that, declaring they felt like schoolboys again, the gentlemen fell on my half dozen pots of preserves in a body, carried them off, and ate them all up then and there, announcing afterwards, there had just been a pot a-piece.

It was really a dreadful time, although we got well cooked plats, for Captain George wasted quite as much as he used. The pigs fed sumptuously that week on his failures, in sauces, minces; puddings, and what not. He had insisted on our making him a paper cap and a linen apron, or rather a dozen linen aprons, for he was perpetually blackening his apron and casting it aside. Then, he used suddenly to cease to take any interest in his occupation, and, seating himself sideways on the kitchen dresser, begin to whistle through a whole opera, or repeat pages of poetry. I tried the experiment of banishing Miss A— — from the kitchen during cooking hours, but a few bars played on the piano were quite enough to distract my cook from his work. My only quiet time was the afternoon, when about four o'clock, my amateur servants all went out for a ride, and left me in peace for a couple of hours. I had enough to do during that short time to tidy up; to collect the scattered books and music, and prepare the tea-supper, for which they came back in tearing spirits, and frantically hungry, between seven and eight o'clock. After this meal had been cleared away, and Mr. U— — and I had washed up (the others declaring they were too tired to stir), we all used to adjourn to the verandah. It happened to be an exceptionally stillweek, no dry, hot nor'-westers, nor cold, wet sou'-westers, and it was perfectly delicious to sit out in the verandah and rest, after the labours of the day, in our cane easy-chairs. The balmy air was so soft and fresh, and the intense silence all around so profound. Unfortunately there was a full moon. I say "unfortunately," because the flood of pale light suggested to these dreadful young men the feasibility of having what they called a "servant's ball." In vain I declared that the housekeeper was never expected to dance. "Oh, yes!" laughed Captain George. "I've often danced with a housekeeper, and very jolly it was too. Come along! F— —, make her dance." And I was forced to gallopade up and down that verandah till I felt half dead with fatigue. The boards had a tremendous spring, and the verandah (built by F— —, by the way), was very wide and roomy, so it made an excellent ball-room. As for the trifling difficulty about music, that was supplied by Captain George and Mr. U— — whistling in turn, time being kept by clapping the top and bottom of my silver butter dish together, cymbal-wise. Oh, dear! It takes my breath away now even to think of those evenings! I see Alice A— — flitting about in her white dress and fern-leaf wreath, dancing like the slender sylph she really was, but never can I forget the odd effect of the gentlemen's feet! No one had their dress boots up at the station, and as Alice and I firmly declined to dance with anybody who wore "Cookham" boots (great heavy things with nails in the soles), they had no other course open to them except to wear their smart slippers. There were slippers of purple velvet, embroidered with gold; others of blue kid, delicately traced in crimson lines; foxes heads stared at us in startling perspective from a scarlet ground; or black jim-crow figures disported themselves on orange tent-stitch. Then these slippers were all more or less of an easy fit, and had a way of flying out on the lawn suddenly, startling my dear dog Nettle out of his first sleep.

Ah, well! that may be an absurd bit of one's life to look back upon, but its days were bright and innocent enough. Health was so perfect that the mere sensation of being alive became happiness, and all the noise of the eager, bustling, pushing world, seemed shut away by those steep hills which folded our quiet valley in their green arms. People have often said to me since, "Surely you would not like to have lived there for ever?" Perhaps not. I can only say that three years of that calm, idyllic life, held no weary hour for me, and I am quite sure that quiet time was a great blessing to me in many ways. First of all, in health, for a person must be in a very bad way indeed for New Zealand air not to do them a world of good; next, in teaching me, amid a great deal of fun and laughter, sundry useful accomplishments, not easily learned in our luxurious civilization; and, lastly, those few years of seclusion from the turmoil of life brought leisure to think out one's own thoughts, and to sift them from other peoples' ideas. Under such circumstances, it is hard if "the unregarded river of our life," as Matthew Arnold so finely call it be not perceived, for one then

"— — Becomes aware of his life's flow

And bears its winding murmur, and he sees

The meadows where it glides, the sun, the breeze;

And there arrives a lull in the hot race,

Wherein he doth for ever chase

That flying and elusive shadow, rest."

One good effect of my sufferings with a house full of unruly volunteers, was that during the brief stay (only two months), of my next cook, I set to work assiduously to learn as many kitchen mysteries as she could teach me, and so became independent of Captain George or F— —, or any other amateur, good, bad, or indifferent.

Nothing could be more extraordinary than the way in which the two affectionate sisters, mentioned [earlier] and who succeeded Euphemia and Lois, quarrelled. They were very unlike each other in appearance, and one fruitful source of bickering arose from their respective styles of beauty. Not only did they wrangle and rave at each other all the day long, during every moment of their spare time, but after they had gone to bed, we could hear them quite plainly calling out to each other from their different rooms. If I begged them to be quiet, there might be silence for a moment, but it would shortly be broken by Maria, calling out, "I say, Dinah, don't you go for to wear green, my girl. I only tell you friendly, but you're a deal too yellow for that. It suits me, 'cause I'm so fresh and rosy, but you never will have my 'plexion, not if you live to be eighty. Good night. I thought I'd just mention it while I remembered." This used to aggravate Dinah dreadfully, and she would retaliate by repeating some complimentary speech of Old Ben's, or Long Tom's, the stockman, and then there would be no peace for an hour.

Their successors were Clarissa and Eunice. Eunice wept sore for a whole month, over her sweeping and cleaning. To this day I have not the dimmest idea why. She gave me warning, amid floods of tears, directly she arrived, though I could not make out any other tangible complaint than that "the dray had jolted as never was;" and to Clarissa, I gave warning the first day I came into the kitchen.

She received me seated on the kitchen table, swinging her legs, which did not nearly touch the floor. She had carefully arranged her position so as to turn her back towards me, and she went on picking her teeth with a hair-pin. I stood aghast at this specimen of colonial manners, which was the more astonishing as I knew the girl had lived in the service of a gentleman's family in the North of England for some time before she sailed.

"Dear me, Clarissa," I cried, "is that the way you behaved at Colonel St. John's?"

Clarissa looked at me very coolly over her shoulder (I must mention she was a very pretty girl, blue-eyed and rosy-cheeked, but with such a temper!) and, giving her plump shoulders a little shrug, said, "No, in course not: they was gentlefolks, they was."

I confess I felt rather nettled at this, and yet it was difficult to be angry with a girl who looked like a grown up and very pretty baby. I restrained my feelings and said, "Well, I should like you to behave here as you did there.

Suppose you get off the table and come and look what we can find in the store room."

"I have looked round," she declared: "there 'aint much to be seen." My patience began to run short, and I said very firmly, "You must get off the table directly, Clarissa, and stand and speak properly; or I shall send you down to Christchurch again." I suppose that was exactly what the damsel wished, for she made no movement; whereat I said in great wrath, "Very well, then you shall leave at the end of a month." And so she did, having bullied everybody out of their lives during that time.

Whilst we are on the subject of manners, it may not be out of place to relate a little episode of my early days "up country." I think I have alluded [in "Station Life in New Zealand"] to our book club; but I don't know that it has been explained that I used to change the books on Sunday afternoon, after our little evening service. It would have been impossible to induce the men to come from an immense distance twice a week, and it was therefore necessary that they should be able to get a fresh book after service. Nothing could have been better than the behaviour of my little congregation: they made it a point of giving no trouble whatever with their horses or dogs, and they were so afraid of being supposed to come for what they could get, that I had some difficulty in inducing those who travelled from a distance to have a cup of tea in the kitchen before they mounted, to set off on their long solitary ride homewards. They were also exceedingly quiet and well-behaved; for if even a dozen men or more were standing outside in fine weather, or waiting within the kitchen if it were wet or windy, not a sound could be heard. If they spoke to each other, it was in the lowest whisper, and they would no more have thought of lighting their pipes anywhere near the house than they would of flying.

This innate tact and true gentlemanly feeling which struck me so much in the labouring man as he appears in New Zealand, made the lapse of good manners, to which I am coming, all the more remarkable. Of course they never touched their hats to me: they would make me a bow or take their hats off, but they never touched them. I have often seen a hand raised involuntarily to the soft felt hat, which every one wears there, but the mechanical action would be arrested by the recollection of the first article of the old colonial creed, "Jack is as good as his master." I never minded this in the least, and got so completely out of the habit of expecting any salutations, that it seemed quite odd to me to receive them again on my return. No, what I objected to was, that when I used to go into my kitchen, about ten minutes or so after the service had been concluded, with the list of club books in my hand, not a single man rose from his seat. They seemed to make it a point to sit down somewhere; on a table or window seat if all the chairs were occupied, but at all events not to be found standing. They would bend their heads and blush, and glance shyly at each other for encouragement as I came in, but no one got up, or took his hat off. This went on for a few weeks, until I felt sure that this curious behaviour did not spring from forgetfulness, or inattention. When I mentioned my grievance in the drawing-room to the gentlemen, I only got laughed at for my pains, and I was asked what else I expected? To this question used to be added sundry anecdotes of earlier colonial life, intended to reconcile me to the manners of these later days. I remember particularly a legend of a man cook, who was said to have walked into the sitting-room of the station where the master was practising tunes on an accordion, and exclaimed, "Now, look here, boss, if you don't leave off that there noise, which perwents me gettin' a wink o' sleep, I'll clear out o' this, sharp, to-morrow mornin'. So now yer know," and with that remark he returned to his bunk.

At last I was goaded to declare I felt sure that the men only behaved in that way from crass ignorance, and that if they knew how much my feelings were hurt, they would alter their manners directly. This opinion was received with such incredulity that I felt roused to declare I should try the experiment next Sunday afternoon. The only warning which at all daunted me was the assurance that I should affront my congregation and scare them away. It was the dread of this which made my heart beat so fast, and my hands turn so cold as I opened the kitchen-door the next Sunday afternoon. There were exactly the same attitudes, every body perfectly civil and respectful, but every body seated. Luckily my courage rose at the right moment, and I came forward as usual with a smile, and said, "Look here, my men, there is one little thing I want to ask you. Do you know that it is not the custom anywhere, in any civilized country, for gentlemen to remain seated and covered when a lady comes into the room? If I were to go into a room in England, where the Prince of Wales, or any of the finest gentlemen of the land were sitting, just as you are now, they would all get up, the Prince first, most likely, and they would certainly take off their hats! Now why can't you all do the same, here?"

The effect of my little speech was magical. Pepper glanced at McQuhair, Moffatt crimsoned and nudged McKenzie, Wiry Ben slipped off the window-seat and shyed his hat across the kitchen, whilst Long Tom, the bullock-driver, "thanked me kindly for mentioning of it;" and every body got up directly and took their hats off. I felt immensely proud of my success, and hastened the moment of my return to the drawing room, where I announced my triumph. I repeated my little speech as concisely as possible; but, alas, it was not nearly so well received as it had been in the kitchen! "Have you ever gone to see a London club?" one person inquired. "Ah: I thought not! I don't know about the Prince, because he always does do the prettiest things at the right moment, but I doubt very much about all the others. I fear you have made a very wild assertion to get your own way." I need hardly say I sulked at that incredulous individual for many days but he always stuck firmly to his own opinion. However, my men never required another hint. They came just as regularly as usual to church, and we all lived happily ever after.

I feel that my chapter should end here; but any record of my New Zealand servants would be incomplete without mention of my "bearded cook." Every body thinks, when I say this, that I am going to tell them about a man, but it is nothing of the sort. Isabella Lyon, in spite of her pronounced beard, was a very fine woman; exceedingly good-humoured looking and fresh-coloured, with most amiable prepossessing manners. She had not long arrived, and had been at once snapped up for an hotel, but she applied for my place, saying she wished for quiet and a country life. Could any thing be more propitious? I thought, like Lois, that my luck, so long in turning, was improving, and that at last I was to have a cook who knew her business. And so she did, thoroughly and delightfully. For one brief fortnight we lived on dainties. Never could I have believed that such a variety of dishes could have been produced out of mutton. In fact we seemed to have everything at table except the staple dish. Unlike the cook who actually sent me in a roast shoulder of mutton for breakfast one morning, Isabella prided herself on eliminating the monotonous animal from her bills of fare. Certainly she was rather heavy on the sauces, etc., and I was trying to pluck up courage to remonstrate, as it would not be easy or cheap to replace them before a certain time of year. And then she was so clean, so smiling, and so good-tempered. She seemed to treat us all as if we were a parcel of children for whom she was never weary of preparing surprises. As for me, I felt miserable if any shepherd or well-to-do handsome young bachelor cockatoo came near the place, dreading lest the wretch should have designs on my cook's heart and hand. I rejoiced in her beard, and would not have had her without it for worlds, as I selfishly hoped it might stand in her matrimonial path.

This Arcadian state of kitchen affairs went on for exactly a fortnight. One evening, at the end of that time, we had been out riding, and returned as usual very hungry. "What are we going to have for supper?" inquired F——. I told him what had been ordered; but when that meal made its appearance, lo, there was not a single dish which I had named! The things were not exactly nasty, but they were queer. For instance, pears are not usually stewed in gravy; but they were by no means bad, and we took it for granted it was something quite new. The housemaid, Sarah, looked very nervous and scared, and glanced at me from time to time with a very wistful look; but I was so delightfully tired and sleepy—one never seemed to get beyond the pleasant stage of those sensations—that I did not ask any questions.

Next morning, when we came out to breakfast, imagine my astonishment at seeing a tureen of half cold soup on the table, and nothing else! I could hardly believe my eyes, and hastened to the kitchen to explain that this was rather too much of a novelty in the gastronomic line. If I live to be a hundred years old, I shall never forget the sight—at once terrible and absurd—which met my eyes. Before the kitchen fire stood Isabella, having evidently slept in her clothes all night. She looked wretched and bloated, and quite curiously dirty, as black as if she had been up the chimney; and even I could see that, early as was the hour, she was hopelessly drunk. Between both of her nerveless, black hands, she held a poker, with which she struck, from time to time, a feeble blow on a piled-up heap of plates, which she persisted in considering a lump of coal. The fire was nearly out, but she hastened to assure me that if she could only break this lump of coal it would soon burn up. Need I say that I rescued my plates at once, and marched the bearded one off to her own apartment.

Oh, how dimmed its dainty freshness had become since even yesterday! Sarah was summoned, and confessed that she had known last night that "Hisabella" had gone on the "burst," having bought, for some fabulous sum, a bottle of rum from a passing swagger. It was all very dreadful, and worst of all was the scene of tears and penitence I had to endure when the rum was finished. The dray, however, relieved me of the incubus of her presence; and that was the only instance of drunkenness I came across among my domestic changes and chances.

Chapter XIV: Our pets.

One of the first things which struck me when I came to know a little more about the feelings and ways of my neighbours in the Malvern Hills, was the good understanding which existed between man and beast. I am afraid I must except the poor sheep, for I never heard them spoken of with affection, nor do I consider that they were the objects of any special humanity even on their owners' parts. This must surely arise from their enormous numbers. "How can you be fond of thousands of anything?" said a shepherd once to me, in answer to some sentimental inquiry of mine respecting his feelings towards his flock. That is the fact. There were too many sheep in our "happy Arcadia" for any body to value or pet them. On a large scale they were looked after carefully. Water, and sheltered feed, and undisturbed camping grounds, all these good things were provided for them, and in return they were expected to yield a large percentage of lambs and a good "clip." Even the touching patience of the poor animals beneath the shears, or amid the dust and noise of the yards, was generally despised as stupidity.

Far different is the feeling of the New Zealander, whether he be squatter or cockatoo, towards his horse and his dog. They are the faithful friends, and often the only companions of the lonely man. Of course there will soon be no "lonely men" anywhere, but a few years ago there were plenty of unwilling Robinson Crusoes in the Middle Island; and whenever I came upon one of these pastoral hermits, I was sure to find a dog or a horse, a cat, or even a hen, established as "mate" to some poor solitary, from whom all human companionship was shut out by mountain, rock, or river.

"Are you not very lonely here?" was often my first instinctive question, as I have dismounted at the door of a shepherd's hut in the back country, and listened to the eternal roar of the river which formed his boundary, or the still more oppressive silence which seemed to have reigned ever since the creation.

"Well, mum, it aint very lively; but I've got Topsy (producing a black kitten from his pocket), and there's the dogs, and I shall have some fowls next year, p'raps."

But my object in beginning this chapter was not to enter into a disquisition on other people's pets, with which after all one can have but a distant acquaintance, but to introduce some of my own especial favourites to those kind and sympathetic readers who take pleasure in hearing of my own somewhat solitary existence in that distant land. I am quite ready to acknowledge that I never thoroughly comprehended the individuality of animals, even of fowls and ducks, until I lived up at the Station. Perhaps, like their masters, they really get to possess more independence of character under those free and easy skies; for where would you meet with such a worldly and selfish cat as "Sandy," or so fastidious and intelligent a smooth terrier as "Rose"? Sandy was an old bachelor of a sleek appearance, red in colour, but with a good deal of white shirt-front and wristbands, as to the get-up of which he was most particular. It was easy to imagine Sandy sitting in a club window; and I am sure he had a slight tendency to gout and reading French novels. Sandy's selfishness was quite open and above-board. He liked you very much until somebody else came whom he liked better, and then he would desert his oldest friend without hesitation. I don't suppose the wildest young colley-pup ever dreamed of chasing or worrying Sandy, who would not have stirred from his warm corner by the fire for Snarleyow himself. Every now and then Sandy must have felt alarmed about his health or his figure, for he ate less, and walked gravely and sulkily up and down the verandah for hours, but as soon as he considered himself out of danger, he relapsed into all his self-indulgent ways. No one ventured to offer Sandy anything but the choicest meats, and he was wont to sit up and beg like a dog for a savoury tit-bit. But he would revenge himself on you afterwards for the humiliation, you might be sure.

What always appeared to me so odd, was that in spite of his known and unblushing selfishness, Sandy used to be a great favourite, and we all vied with each other for the honour of his notice. Now why was this? If boundless time and space were at our disposal, we might go deeply into the question and work it out, but as the dimensions of this volume are not elastic, the impending social essay shall be postponed, and we will confine ourselves to a brief description of Sandy's outer cat. He was of a pure breed, far removed from the long-legged, lanky race of ordinary station-cats, who from time to time disappeared into the bush and contracted alliances with the still more degraded specimens of their class who had long been wild among the scrub. No: Sandy came of "pur sang," and held his small square head erect, with a haughty carriage as beseemed his ancestry. His fur was really beautiful, a sort of tortoiseshell red, the lighter stripes repeating exactly the different golden tints of a fashionable chignon. In early youth, though it is difficult to imagine Sandy ever a playful kitten, his tail had been curtailed to the length of three inches, and this short, flexible stump gave an air of great decision to Sandy's movements. But his chief peculiarity, and I must add, attraction, in my opinion, was the perfume of his sleek coat. When Sandy condescended to take his evening doze on my linsey lap, I never smelt anything so strange and so agreeable as the odour of his fur, specially that on the top of his head. It was like the most delicate musk, but without any of the sickly smell common to that scent. I believe Sandy knew of this personal peculiarity, and felt proud of it.

A far more unselfish and agreeable personage was Rose, the white terrier, whose name often finds a loving place in these pages. She and Sandy dwelt together in peace and amity, although the little doggie never could have felt any affection for her selfish companion. Rose's nerves were of a delicate and high-strung order, and there was nothing she hated so much as uproarious noise. Every now and then it chanced that during a few days of wet or windy weather, our little house had been filled by passing guests: gentlemen who had called in to ask for supper and a bed, intending to go on next day. In a country where inns or accommodation-houses are fifty miles apart, this is a common incident, and it sometimes happens that the resources of station hospitality are taxed to the utmost in this way. I have known our own little wooden box to be so closely packed, that besides a guest on each sofa in the drawing-room, there would be another on a sort of portable couch in the dining-room. This was after the spare room had been filled to the utmost. A delicate "new chum," who required to be pampered, had retired to rest on the hard kitchen sofa described elsewhere; whilst a couple of sturdy travellers were sleeping soundly in the saddle room. After that, there could be nothing for the last comer except a shake-down in red blankets.

It always happened I observed that everybody arrived together. For weeks we would be alone. I lived once for eight months without seeing a lady; and then, some fine evening, half a dozen acquaintances would "turn up,"—there really is no other word for it. Well, on these occasions, when, instead of departing next morning, our impromptu guests have sometimes been forced to wait until such time as the rain or the wind should cease; their pent-up animal spirits became often too much for them, and they would feel an irresistible impulse to get rid of some of their superfluous health and strength by violent exercise. I set my face at once against "athletic sports" or "feats of strength" being performed in my little drawing-room, although they were always very anxious to secure me for the solitary spectator; and I forget who hit upon the happy thought of turning the empty wool-shed into a temporary gymnasium. There these wild boys—for, in spite of stalwart frames and bushy beards, the Southern Colonist's heart keeps very fresh and young—used to adjourn, and hop and leap, wrestle and box, fence and spar, to their active young limbs' content. They seemed very happy, and loud were the joyous shouts and peals of laughter over the failures; but after seeing the performance once or twice, I generally became tired and bored, and used to slip away to the house and my quiet corner by the fire. Rose considered it her duty to remain at her master's heels as long as possible, but after a time she too would creep back to silence and warmth, though she never deserted her post until the noise grew altogether too much for her nerves; and then, with a despairing whimper, sometimes swelling to a howl, poor little Rose would tuck her tail between her legs, and dash out, through the storm, to seek shelter and quiet with me.

Whenever Rose appeared thus suddenly in my quiet retreat, I felt sure some greater uproar than usual was going on down at the wool-shed, and, more than once, on inquiry, I found Rose's nerves must have been tried to the utmost before she turned and fled.

As for the intelligence of sheep-dogs, a volume could be written on the facts concerning them, and a still more entertaining book on the fictions, for a New Zealand shepherd will always consider it a point of honour to cap his neighbour's anecdote of his dog's sagacity, by a yet stronger proof of canine intelligence. I shall only, briefly allude to one dog, whose history will probably be placed in the colonial archives,—a colley, who knows his master's brand; and who will, when the sheep get boxed, that is mixed together, pick out; with unfailing accuracy, all the bleating members of his own flock from amid the confused, terrified mass. As for the patience of a good dog in crossing sheep over a river, I have witnessed that myself, and been forced to draw conclusions very much in favour of the dog over the human beings who were directing the operation. Some dogs again, who are perfectly helpless with sheep, are unrivalled with cattle, and I have stood on the edge of a swamp more than once, and seen a dog go after a couple of milch cows, and fetch them out of a herd of bullocks, returning for the second "milky mother" after the first had been brought right up within reach of the stockman's lash.

Then among my horse friends was a certain Suffolk "Punch," who had been christened the "Artful Dodger," from his trick of counterfeiting lameness the moment he was put in the shafts of a dray. That is to say if the dray was loaded; so long as it was empty, or the load was light, the "Dodger" stepped out gaily, but if he found the dray at all heavy, he affected to fall dead lame. The old strain of staunch blood was too strong in his veins to allow him to refuse or jib, or stand still. Oh, no! The "Dodger" arranged a compromise with his conscience, and though he pulled manfully, he resorted to this lazy subterfuge. More than once with a "new chum" it had succeeded to perfection, and the "Dodger" found himself back again in his stable with a rack of hay before him, whilst his deluded owner or driver was running all over the place to find a substitute in the shafts. If I had not seen it myself, I could not have believed it. In order to induce the "Dodger" to act his part thoroughly, a drayman was appointed whom the horse had never seen, and therefore imagined could be easily imposed upon. The moment the signal was given to start, the "Dodger," after a glance round, which plainly said, "I wonder if I may try it upon you," took a step forward and almost fell down, so desperate was his lameness. The driver, who was well instructed in his part, ran round, and lifted up one sturdy bay leg after the other, with every appearance of the deepest concern. This encouraged the "Dodger," who uttered a groan, but still seemed determined to do his best, and limped and stumbled a yard or two further on. I confess it seemed impossible to believe the horse to be quite sound, and if it had depended on me, the "Dodger" would instantly have been unharnessed and put back in his stable. But the moment had come to unmask him. His master stepped forward, and pulling first one cunning ear, on the alert for every word, and then the other; cried, "It wont do, sir! step out directly, and don't let us have any nonsense." The "Dodger" groaned again, this time from his heart probably, shook himself, and, leaning well forward in his big collar, stepped out without a murmur. The lameness had disappeared by magic, nor was there even the slightest return of it until he saw a new driver, and considered it safe to try his oft-successful "dodge" once more.

Very different was "Star," poor, wilful, beauty, whose name and fate will long be remembered among the green hills, where her short life was passed. Born and bred on the station, she was the pride and joy of her owner's heart. Slender without being weedy, compact without clumsiness, her small head well set on her graceful neck, and her fine legs, with their sinews like steel, she attracted the envy of all the neighbouring squatters. "What will you take for that little grey filly when she is broken?" was a constant question. "She's not for sale," her owner used to answer. "I'll break her myself, and make her as gentle as a dog, and she'll do for my wife when I get one." But this proved a castle in the air, so far as Star was concerned. The wife was not so mythical. In due time she appeared in that sheltered valley, and, standing at the head of a mound marked by a stake whereon a star was rudely carved, heard the story of the poor creature's fate. From the first week of her life, Star (so-called from a black, five-pointed mark on her forehead), showed signs of possessing a strange wild nature. Unlike her sire or dam, she evidently had a violent temper,—and not to put too fine a point on it,—was as vicious a grey mare as ever flung up her heels in a New Zealand valley.

When her second birthday was passed, Star's education commenced. The process called "gentling," was a complete misnomer for the series of buck jumps, of bites and kicks, with which the young lady received the slightest attempt to touch her. She had a horrible habit also of shrieking, really almost like a human being in a frantic rage; she would rush at you with a wild scream of fury, and after striking at you with her front hoofs, would wheel round like lightning, and dash her hind legs in your face. The stoutest stockman declined to have anything whatever to do with Star; the most experienced breaker "declined her, with thanks;" generally adding a long bill for repairs of rack and manger, and breaking tackle, and not unfrequently a hospital report of maimed and wounded stablemen. Amateur horsemen of celebrity arrived at the station to look at the beautiful fiend, and departed, saying they would rather not have anything to say to her. At last, she was given over in despair, to lead her own free life, never having endured the indignity of bit or bridle for more than two minutes.

Months passed away, and Star and her tantrums had been nearly forgotten, when one mild winter evening the stockman came in to report that,—wonder of wonders,—Star was standing meekly outside, whinnying, and as "quiet as a dog." Her master went out to find the man's report exact: Star walked straight up to him, and rubbed her soft nose confidingly against his sleeve. The mystery explained itself at a glance: she was on the point of having her first foal, and, with some strange and pathetic instinct, she bethought herself of the kind hands whose caresses she had so often rejected, and came straight to them for help and succour. Her shy and touching advances were warmly responded to, and in a few minutes the poor beast was safely housed in the warm shed which then represented the present row of neat stables long since on that very spot. A warm mash was eagerly swallowed, and the good-hearted stockman volunteered to remain up until all should be happily over; but his courage failed him at the sight of her horrible sufferings, and in the early dawn he came to rouse up his master, and beg him to come and see if anything more could be done. There lay Star, all her fierce spirit quenched, with an appealing look in her large black eyes, which seemed positively human in their capacity for expressing suffering. It was many hours before a dead foal was born, and there is no doubt that if she had been out on the bleak hills, the poor exhausted young mother must have perished from weakness. She appeared to understand thoroughly the motive of all that was being done for her, and submitted with patience to all the remedies. Gradually, but slowly, her strength returned; and, alas, her evil nature, tamed by anguish, returned also! Day by day she became shyer of even the hand which had fed and succoured her; and, as this is a true chronicle, it must be stated that the very first use Mrs. Star made of her convalescence was, to kick her nurse on the leg, break her halter into fragments, and gallop off to the hills with a loud neigh of defiance. Whenever the topic of feminine ingratitude came on the carpet at that station, this, which Star had done, used always to be told as an instance in point.

Two years later, exactly the same thing happened again. The dreaded hour of suffering found the wayward beauty once more under the roof which had sheltered her in her former time of trial, and once more she rested her head in penitence and appeal against her owner's shoulder. Who could bear malice in the presence of such dreadful pain? Not Star's owner, certainly. Besides the home resources, a man on horseback was sent off to fetch a famous veterinary who chanced to be staying at a neighbouring station, and they both returned before Star's worst sufferings began. All that skill and experience could do was done that night; but the morning light found the poor little grey mare dying from exhaustion, with another dead foal lying by her side. She only lived a few hours later, in spite of stimulants and the utmost care, and died gently and peacefully, with those human hands whose lightest touch she had so flouted, ministering tenderly to her great needs. The stockman had become so fond of the wayward beauty, in spite of her ingratitude, that the only solace he could find for his regret at her early death, lay in digging a deep grave for her, and carving the emblem of her pretty name on the rude stake which still marks the spot.

No account of station pets would be complete without a brief allusion to my numerous and unsuccessful attempts to rear merino lambs in the house. It never was of any use advising me to leave the poor little creatures out on the bleak hill-side, if, in the course of my rambles after ferns or creepers, I came upon a dead ewe with her half-starved baby running round and round her. How could I turn my back on the little orphan, who, instead of bounding off up the steep hill, used to run confidingly up to me, and poke its black muzzle into my hand, as if it would say, "Here is a friend at last"? And then merino lambs are so much prettier than any I have seen in England. Their snow-white wool is as tightly screwed up in small curls as any Astracan fleece, and from being of so much more active a race, they are smaller and more compact than English lambs, and not so awkward and leggy. A merino lamb of a couple of hours old is far better fitted to take care of itself up a mountain than a civilized and helpless lamb of a month old, besides these latter being so weak about the knees always. I only mention this, not out of any desire to "blow" about our sheep, but because I want to account for my tender-heartedness on the subject of desolate orphans. The ewes scarcely ever died of disease, unless by a rare chance it happened to be a very old lady whose constitution gave way at last before a severe winter. We oftenest found that the dead mother was a fine fat young ewe; who had slipped up on a hill-side and could not recover herself, but had died of exhaustion and fatigue from her violent efforts to kick herself up again. If we chanced to be in time to rescue her by the simple process of setting her on her legs again, it would be all right, but sometimes the poor creature had been cold and stiff for hours before we found her, and her lamb had bleated itself hoarse and hungry, and was as tame as a pet dog. Now who could turn away from a little helpless thing like that, who positively leaped into your arms and cuddled itself up in delight, sucking vigorously away at your glove, or anything handy? Not I, for one,—though I might as well have left it alone, so far as its ultimate fate was concerned; but I always hoped for better luck next time, and carried it off in my arms.

The first thing to be do be on arrival at home, was to give the starving little creature a good meal out of a tea-pot, and the next, to put it to sleep in a box of hay in a warm corner of the kitchen. What always seemed to me so extraordinary, was that the lambs, one and all, preserved the most cheerful demeanour, ate and drank and slept well,—and yet died within a month. Some lingered until quite four weeks had passed, others succumbed to my treatment in a week. I varied their food, mixing oatmeal with the milk;

some I fed often, others seldom; to some I gave sugar in the milk, others had new milk. There was abundance of grass just outside the house for them to eat, if they could. Some did mumble feebly at it, I remember, but the mortality continued uninterrupted. It must have been very ridiculous to a visitor, to see my dear little snowy pets going down on their front knees before me, and wagging their long tails furiously the moment the tea-pot was brought out. They were far too sensible to do this if my hands were empty. Gentle, affectionate little creatures, they used to be wonderfully well-behaved, though now and then they would wander through the verandah, and so into my bedroom, where the drapery of my dressing-table afforded them endless amusement and occupation. They gnawed and sucked all my "daisy" fringe, until the first thing that had to be done when a lamb arrived at the house, was to take off muslins and fringes from that, the only trimmed table in the house.

Often and often, of a cold night (for we must remember that New Zealand lambing used always to come off in winter), we would all become suddenly aware of a strong smell of burning pervading the whole house; which, on being traced to its source, was often found to proceed from the rosette of wool on the forehead of a chilly lamb. The creature drew nearer and nearer to the genial warmth of the kitchen fire, until at last it used to lean its brow pensively against the red hot bars. Hence arose the powerful odour gradually filling the whole of the little wooden house. Of course I used to rush to the rescue, and draw my bewildered pet away from the fatal warmth, but not until it had usually singed the wool off down to the bone, and there was often a bad burn on its forehead as well. But still, in spite of stupidity and an insatiable appetite, I always grieved very sincerely for each of my orphan lambs as it in turn sank into its early grave. I used to be well laughed at for attaching any sentiment to an animal which had sunk so disgracefully low in the money-market as a New Zealand lamb, but the abundant supply of my little pets never made it easier for me to lose the particular one which I had set my heart on rearing. It certainly did afford me some comfort to hear that merino lambs had always been difficult, if not impossible to bring up, like so many "pups," by hand; and among all the statistics I carefully collected, I could only find one well-authenticated instance of a foundling having been reared indoors. My informant tried to comfort me by tales of the tyranny that stout and tame sheep exercised over the household which had sheltered it, but I fear that the stories of its delightful impudence only made me more anxious to succeed in my own baby-farming experiments among the lambs.

Chapter XV: A feathered pet.

No record of those dear, distant days would be complete without a short memoir of "Kitty." She was only a grey Dorking hen, but no heroine in fact or fiction, no Lady Rachel Russell or Fleurange, ever exceeded Kitty in unswerving devotion to a beloved object, or rather objects.

To see Kitty was to admire her, at least as I saw her one beautiful spring evening in a grassy paddock on the banks of the Horarata. We had ridden over there to visit our kind and friendly neighbours, the C— —'s; we had enjoyed a delicious cup of tea in the passion-flower-covered verandah, which looked on the whole range, from East to West, of the glorious Southern Alps, their shining white summits sharply cut against our own peculiarly beautiful sky; we had strolled round the charming, unformal garden, on either sloping side of a wide creek, and had admired, with just a tinge of envy, the fruits and flowers, the standard apple and rose trees, the tangle of fern and creepers, the wealth of the old and new worlds heaped together in floral profusion; we had done all this, I say, and very pleasant we had found it. Now we were trying to say goodbye: not so easy a task, let me tell you, when there are so many temptations to linger, and when you are greatly pressed to stay. The last device of our hospitable hostess to keep us consisted in offering to show me her poultry-yard. Now I was a young beginner in that line myself, and tormented my ducks and fowls to death by my incessant care: at least that is the conclusion I have arrived at since; but at that time, I considered it as necessary to look after them as if they had been so many children. The consequence was,—as I pathetically complained to Mrs. C— —, that my hens sat furiously for a week, and then took to lingering outside, where perpetual feeding was going on, until their eggs grew cold; that my ducks neglected their offspring and allowed the rats to decimate them, and that every variety of epidemic and misfortune assailed in turns my unhappy poultry yard. Kind Mrs. C— — listened as gravely as she could, hinting very gently, that perhaps I took too much trouble about them; then, fearing least she might have wounded my feelings, she hastened to suggest that I should try the introduction of a different breed.

As a preliminary step to this reformation, she offered to bestow upon me one of her best Dorking hens. It was too tempting an offer to be refused, and I forthwith bestowed my affections on a beautiful grey pullet, whose dignified carriage and speckled exterior bespoke her high lineage. "That's Kitty," said Mrs. C— —. "I am so glad you fancy her; she is one of my nicest young hens. We'll catch her for you in a moment." I must pause to mention here, that it struck me as being very odd in New Zealand the way in which every creature has a name, excepting always the poor sheep. If one sees a cock strutting proudly outside a shepherd's door; you are sure to hear it is either Nelson or Wellington; every hen has a pet name, and answers to it; so have the ducks and geese,—at least, up-country; of course, dogs, horses, cows and bullocks, each rejoice in the most inflated appellations, but I don't remember ever hearing ducks and fowls answer to their names in any other country.

But this is only by the way. I gratefully and gladly accepted the transfer of the fair Kitty, and only wondered how I was to convey her to her new home, fifteen miles away. Kitty was soon caught, and carried off into the house to be packed up for her first ride. Accustomed as I am to ridiculous things happening to me, still I never felt in so absurd a position as when, having mounted "Helen," who seemed in a particularly playful mood after a good feed of oats, Kitty was handed to me neatly tied up in a pillow-case with her tufted head protruding from a hole in the seam at the side. Although very anxious to carry her home immediately, my heart died within me at the prospect of a long gallop on a skittish mare with a plump Dorking hen tied up in a bag on my lap.

There was no help for it, however, and I tried to put my bravest face on the matter. The difficulties commenced at the very point of departure, for it is not easy to say farewell cordially with your hands full of reins, whip, and poultry. But it proved comparatively easy going whilst we only cantered over the plains. It was not until the first creek had been reached, that I really perceived what lay before me. Helen distrusted the contents of the bag, and kept trying to look round and see what it contained; and her fears of something uncanny might well have been confirmed when she took off at her first flat jump. Kitty screamed, or shrieked, or whatever name best expresses her discordant and piercing yells. I more than suspect I shrieked too, partly at the difficulty of keeping both Kitty and Helen in any sort of order, and partly at my own insecurity. No sooner had Helen landed on the other side, than she fled homewards as if a tin kettle were tied to her tail. The speed at which we dashed through the fragrant summer air completely took away Kitty's breath, and the poor creature appeared more dead than alive by the time I dismounted, trembling myself in every limb for her safety as well as my own, at the garden gate.

However, next morning brought a renewed delight in existence to both Kitty and me, and our night's sleep had made us forget our agitation and peril. After breakfast I introduced her to the poultry yard, and she adapted herself to her new home with a tact and good humour most edifying to behold. Months passed away. Kitty had made herself a nest in a place, the selection of which did equal honour to her head and heart, and she gladdened my eyes one fine morning by appearing with a lovely brood of chicks around her. Who so proud as the young mother? She exhibited them to me, and after I had duly admired them, used to carry them off to a nursery of her own, which she had established among the tussocks just outside the stable door. Mrs. C—— had impressed upon me that Kitty could be safely trusted to manage her own affairs. No fear of her dragging her fluffy babies out among the wet grass too early in the morning, or losing them among the flax bushes on the hill-side. No: Kitty came of a race who were model mothers, and was to be left to take care of herself and her chickens.

About a week after Kitty had first shown me her large, small family, a friend of ours arrived unexpectedly to stop the night. Next morning, when he was going away, he apologised for asking leave to mount at the stables, saying his led horse was so vicious, and the one he was riding so gay, that it was quite possible their legs might find themselves within the verandah, or do some mischief to the young shrubs which were the pride and joy of my heart. This gentleman rode beautifully, and I used to like to see the courage and patience with which he always conquered the most unruly horse.

"We will come up to the stable and see you mount," I cried, seizing my hat. Of course every one followed my lead, and it was to the sound of mingled jeers and compliments that poor Mr. T— — mounted his fiery steed, and seized hold of the leading rein of his pack-horse. But this animal had no intention of taking his departure with propriety or tranquillity: he pranced and shied, flinging out his heels as he wildly danced round to every point of the compass, in a circle. Gradually he drew Mr. T— — and his chestnut a dozen yards away from the stable, and it was just then that I perceived poor Kitty sitting close under a tussock. It chanced to be the hour for the chickens' siesta, and they were all folded away beneath her ample brooding wings. Perhaps the danger had come too near to be avoided before I perceived it, but at all events my loud shriek of warning was too late to save the pretty crouching head from the flourish of the pack-horse's glancing heels. Swift indeed was the blow; for scarcely ten seconds could have passed between my first glimpse of poor Kitty's bright black eye looking out, with such mortal terror in its expression, from beneath the yellow tuft of grass, and my seeing the horse's heel lay her head right open. The brave little mother never dreamed of saving herself at the cost of her nestlings. She crouched as low as possible, and when the horse had jumped over her I flew to see if she had escaped. No. There lay my pretty pet, with her wings still outspread and her chickens unhurt. But she seemed dead: her head had been actually cut clean open, and I never expected that she would have lived a moment. Yet she did. I took her at once to the well hard by, and bound up her split head with my pocket handkerchief, keeping it well wetted with cold water. Later on I put forth all the surgical art I possessed, and dressed the wound in the most scientific manner, nursing poor Kitty tenderly in the kitchen, and feeding her with my own hands every two hours. She was for a long time incapable of feeding herself and; even when all danger was over, required most careful nursing. However, the end of the story is that, she recovered entirely her bodily health, but her poor little brain remained clouded for ever. She never took any more notice of her chickens, who had to be brought up by hand, and she never mixed again with the society of the poultry-yard. At night she roosted apart in the coalshed, and she never seemed to hear my voice or distinguish me from others, though she was perfectly tame to everybody. Kitty's end was very tragical. She grew exceedingly fat, and at last, one time when we were all snowed up and could not afford to be sentimental, my cook laid hold of poor Kitty, who was moping in her usual corner, and converted her into a savoury stew without telling me, until I had actually dined off her. I was very angry; but Eliza only repeated by way of consolation, "She had no wits, only flesh, consequently she was better in my stew-pot nor anywhere else, mum, if you'll only look at it calm like." But it was very hard to be made to eat one's patient, especially when I was so proud of the way her poor head had healed.

If anybody wanted to teaze me, they suggested that I had omitted to replace my dear Kitty's brains before closing that cruel wound in her skull.

Chapter XVI: Doctoring without a diploma.

So many reminiscences come crowding into my mind,—some grave and others gay,—as I sit down to write these final chapters, that I hardly know where to begin.

The most clamorous of the fast-thronging memories, the one which pushes its way most vividly to the front, is of a little amateur doctoring of mine; and as my patient luckily did not die of my remedies, I need not fear that I shall be asked for my diploma.

Shearing was just over; over only that very evening in fact. We had been leading a sort of uncomfortable picnic life at the home station for more than ten days, and had returned to our own pretty little home up the valley, late on Saturday night, in time for the supper-dinner I have so often described. It was my doing, that fortnight's picnic at the home station, and I may as well candidly confess it was a mistake; although, made, like most mistakes in life, with good intentions. Our partner had gone to England, our manager had just left us to set up sheep-farming on his own account, and all the responsibility of shearing a good many thousand sheep devolved on F— —. And not only the shearing; the flock had to be carefully draughted, the ewes, wethers, and hoggets, to be branded, ear-marked, and turned out on their several ranges; the wethers for home consumption, which consisted of a good-sized flock of many hundred sheep, turned into the home-paddock,—an enclosure of some five or six hundred acres,—and various other minute details to be seen to; the wool to be sent down to Christchurch, and the stores brought up by the return drays.

My motives for the plan I formed for us to go over, bag and baggage, to the home station, the evening before the shearing began, and live there till it was over, were varied. We will put the most unselfish first, for the sake of appearances. I knew it would be very hard work for poor F— — all that time, and I thought it would add to his fatigue if he had to go backwards and forwards to his own house every day, getting up at five in the morning and returning late at night, besides having no comfortable meals. The next motive was that I wanted very much to see the whole process of shearing, and all the rest of it, myself; and as it turned out, though I little dreamed of it at the time, this proved to be my only chance. Every body tried to dissuade me from carrying out the scheme, by urging that I should be very uncomfortable; but I did not care in the least for that, and insisted on being allowed at all events to see how I liked it.

Accordingly one evening we set forth: such a ridiculous cavalcade. I would not hear of riding, for it was only a short two miles walk; and as we did not start until after our last meal, the sun had dipped behind Flag-pole's tall peak, and nearly the whole of our happy valley lay in deep, cool shadow. Besides which, it looked more like the real thing to walk, and that was half the battle with me. The "real thing" in this case, though I did not stop to explain it to myself, must have meant emigrants, Mormons, soldiers on the march, what you will; any thing which expresses all one's belongings being packed into a little cart, with a huge tin bath secured on the top of all. Such a miscellaneous assortment of dry goods as that cart held! A couple of mattresses (for my courage failed me at the idea of sleeping on chopped tussocks for a fortnight), a couple of folding-up arm-chairs, though, as it turned out, one would have been enough, for poor F— — never sat down from the time he got up until he went to bed again; a large hamper of provisions, some books, our clothes, and various little matters which were indispensable if one had to live in an empty house for a fortnight. I had sent my two maids over one morning a few days before, with pails and mops and brushes, and they had given the couple of rooms which we were to inhabit, a thorough good cleaning and scouring, so my mind was easy on that point. It would not have answered, for many reasons, to have encumbered ourselves with these damsels during our stay at the home station. In the first place, there was really no accommodation for them; in the next, it would have entailed more luggage than the little cart could hold; and, finally, we should have been obliged to leave them behind at the last moment: for only the evening before we started, a couple of friends arrived, in true New Zealand fashion, from Christchurch, to pay us a month's visit. It was too late to alter our plans then, so we told them to, make themselves thoroughly at home, and took our departure next day in the way I have alluded to.

We had plenty of escort as far as the first swamp. When that treacherous and well-known spot had been reached, everybody suddenly remembered that they had forgotten something or the other which obliged them to return directly, so our farewells had to be exchanged from the centre of a flax bush. The cart meanwhile was nearly out of sight, so wide a detour had its driver been forced to make in order to find a place sound enough to bear its weight. But we caught it up again after we had happily crossed the quagmire which used always to be my bug-bear, and in due time we made our appearance, in the gloaming, at the tiny house belonging to the home station. Early as was the hour, not later than half-past eight, the place lay silent and still under the balmy summer haze. All the shearers were fast asleep in the men's hut, whilst every available nook and corner was filled with the spare hands; the musterers, branders, yard-keepers, and many others, whose duties were less-defined. Far down the flat we could dimly discern a white patch,—the fleecy outlines of the large mob destined to fill the skillions at day-break to-morrow morning; and, although we could not see them distinctly, close by, watchful and vigilant all through that and many subsequent summer nights, Pepper and his two beautiful colleys kept watch and ward over the sheep.

Writing in the heavy atmosphere of this vast London world, I look back upon that, and such evenings as that, with a desperate craving to breathe once more he delicious air unsoiled by human lungs, and stirred into fresh fragrance by every summer sigh of those distant New Zealand valleys. No wonder people were always well in such a pure, clear, light atmosphere. I try to feel again in fancy the exquisite enjoyment of merely drawing a deep breath, the thrilling sensation of health and strength it sent tingling down to your finger ends. No fleck or film of vapour or miasma could be seen or smelt, though the day had been burning hot, and, as I have said, there were plenty of creeks and swamps hard by. Damp is unknown in those valleys, and we might have lingered bareheaded even after the heavy dew began to fall, without risk of cold, or fever, or any other ailment. But we could not afford to linger a moment out of doors that lovely tempting evening. F— — and the driver of the cart, who had some important part to take in the morrow's proceedings (I forget exactly what), soon tossed out my little stores, which looked very insignificant as they lay in a heap in the verandah, and departed to see that all was in train for next day's work. I had no time to enjoy the evening's soft beauty: the beds had to be made; clothes to be unpacked and hung up; stores must be arranged on the shelves in the sitting-room,—for the house only consisted of two small rooms in front, with a wide verandah, and a sort of lean-to at the back, which was divided into a small kitchen and store-room. This last was empty. I confess I thought rather regretfully of my pretty, comfortable, English-looking bed-room at the other house, with its curtains and carpet, its wardrobes and looking-glasses, when I found myself surveying the scene of my completed labours. Two station bunks,—i.e., wooden bed-frames of the simplest and rudest construction, with a sacking bottom,—a couple of empty boxes, one for a dressing-table and the other for a wash-stand, a tin basin and a bucket of water, being the paraphernalia of the latter, whilst some nails behind the door served to hang our clothes on, such was my station bedroom and all my own doing too! Certainly it looked uncomfortable enough to satisfy any one, but I would not have complained of it for the world, lest I might have been ordered home directly.

Hard as was my bed that night, I slept soundly, and it appeared only five minutes before I heard a tremendous noise outside the verandah. The bleating of hundreds of sheep announced that the mob were slowly advancing, before a perfect army of men and dogs, up to the sheep yards. What a din they all made! F—— was wide awake, and up in a moment. I, anxious to show why I had insisted on coming over, got up too, and made my way into the little kitchen, where I found a charming surprise awaiting me in the shape of some faggots of neatly-stacked wood, cut into exactly, the right lengths for the American stove; and also a heap of dry Menuka bushes, which make the best touchwood for lighting fires in the whole world. The tiny kitchen and stove were both scrupulously clean, and so were my three saucepans and kettle. This had been, of course, my maids' doing, but the fuel was a delicate little attention on Pepper's part. How he blushed and grinned with delight when I thanked him before all his mates! This was indeed station-life made easy! It did not take two minutes to light my fire, and in five more I had a delicious cup of tea and some bread-and-butter all ready for F— —. It was nearly cold, however, by the time I could catch him and make him drink it. Of course, being a man, instead of saying, "Thank you," or anything of that sort, he merely remarked, "What nonsense!" but equally of course, he was very glad to get it, and ate and drank it all up, returning instantly to his shed.

After this little episode, I set to work to unpack a little, and make the sitting-room look the least bit more home-like; then I laid the cloth for breakfast, put out the pie and potted meat, etc. (no words can say how heartily tired of pies we both were before the week was over), and arranged everything for breakfast. Then I waylaid one of the numerous stray "hands" which hang about a station at shearing time, and got him to fetch me a couple of buckets of water as far as the verandah. These I conveyed myself into the little sleeping-room, and finished my toilette at my leisure: tidying it all up afterwards. I wonder if any one has any idea what hot work it is making a bed? So hot, in fact, that I resolved in future to be wise enough to finish all these domestic occupations before I had my bath. The worst of getting up so early proved to be that by nine o'clock I was very tired, and had nothing else to do for the remainder of the long, noisy day. As for the meals, they were wretchedly unsociable; for F— — only came in to snatch a mouthful or two, standing, and it was of little use trying to make things comfortable for him. I must confess here, what I would not acknowledge at the time, that I found it a very long and dull visit. My husband never had time to speak to me, and when he did, it was only about sheep. I grew weary of living on cold meat, for it was really too hot to cook; and my servants used to send me over, every second day, cold fowls or pies; besides, one seemed to live in a whirl and confusion of dust, and bleating, and barking. After the day's work was fairly over, F— — used to rush in, seize a big bath-towel, cry "I am off for a bathe in the creek," and only return in time for supper and bed. The weather was all that a sheep-farmer could desire. Bright, sunny, and clear, one lovely summer day followed another; hot, almost to tropical warmth, without any risk or fear of sun-stroke or head-ache, and a delicious lightness in the atmosphere all the time, which merged into a cool bracing air the moment the sun had slowly travelled behind the high hills to the westward.

But all these details, though necessary to make you understand what I had been doing, are not the story itself, so to that we will hurry on. The shearing was over; Saturday evening had come, as welcome to poor imprisoned me as to any one, and the great work of the New Zealand year had been most successfully accomplished. F— — was in such good humour that he even deigned to admit that his own comfort had been somewhat increased by my living at the home station, so I felt quite rewarded for my many dreary hours. The shearers had been paid, and were even then picking their way over the hills in little groups of two and three; some, I grieve to say, bound for the nearest accommodation-house or wayside inn, and others for the next station, across the river, where the skillions were full, and waiting for them to begin on Monday morning. Only half-a-dozen people, instead of thirty, were left at our place, and there would not even have been so many if it had not been thought well to keep a few there until the bale-loft was empty. Generally it was arranged for the wool-drays to follow each other every two days with a load down to Christchurch; for the greatest risk a sheep-farmer runs is from his shed taking fire whilst it is full of bales of wool. This had happened often enough in the colony, and even in our neighbourhood, to make us more and more careful every year; and, as I have said, amongst our precautions, was that of keeping as little wool as possible in the shed. Most flock-owners waited until the shearing should be quite over before they carted the wool away; but in that case, a spark from a pipe, a match carelessly dropped in a tussock outside, when a nor'-wester was blowing,—and the slight wooden building would be blazing like a torch, and your year's income vanishing in the smoke!

Even at the last moment, when the cart had already started homewards, with the tin bath balanced once more on the top of the mattresses and boxes; when the house was empty, and I was waiting, my hat and jacket on, and flax-stick in hand, eager to set out, a doubt arose about the expediency of our return home. Some accidental delay had prevented the dray from arriving in time to start for Christchurch with the last load, and between two and three hundred pounds worth of wool still remained in the shed,—packed and labelled indeed, but neither insured nor protected from the risk of fire in any way. F— — was very loath to leave them there; but, yielding to my entreaties, he called Pepper, the head shepherd, and solemnly gave the wool-shed and its contents over into his charge, with many and many a caution about fire. Pepper was as trustworthy and steady a shepherd as any in the colony, and promised to "keep his weather-eye open," as he phrased it, in nautical slang picked up from some run-away sailor.

All the way home F— — said from time to time, anxiously, "I wish the shed was empty;" but I cheered him up, and told him he was over-tired and unreasonably nervous, and so forth, but with a great longing myself for Monday morning to come, and for the dray to take its load and start. I need not dwell on how delicious it was to return home, where everything seemed so comfortable and nice, and the bed felt especially soft and welcome to tired limbs. Early were our hours, you may be sure, and we slept the sleep of the hard-worked until between two and three o'clock the next morning. Then we were roused up by some one knocking loudly against our wide-open latticed window.

I was the first to hear the noise, and cried, "Who's there? what is it?" all in a breath.

"The wool-shed on fire," murmured F— —, in a tone of agonized conviction.

"It's you that's wanted, please mum, this moment, over at the home station!" I heard Pepper say, in impatient tones.

"It's the wool-shed," repeated F— —, more than half asleep, and with only room for that one idea in his dreamy mind.

"Nonsense!" I cried, jumping out of bed. "I should not be wanted if the wool-shed were on fire. Don't you hear Pepper say he wants me?"

"All right, then," said F— —, actually turning over and proposing to go to sleep again. But there was no more sleep for either of us that night. Whilst I hastily put on my riding-habit, Pepper told me, through the window; an incoherent tale of some one being at the point of death, and wanting me to cure him, and the master to bring over pen and ink, to make a will, and dying speeches and cold shivers, all mixed up together in a tangle of words. F—— took some minutes to understand that it was Fenwick, a gigantic Yorkshireman, who had been seized with what Pepper would call the "choleraics," and who, in spite of having swallowed all the mustard and rum and "pain-killer" left on the premises, grew worse and worse every moment. "He's dying, safe enough," concluded Pepper, "but he's main anxious to see you, mum, and the master; and he wants a Bible brought to swear him, and he's powerful uneasy to make his will." I knew quite as little of medicine as my husband did of law, but of course we decided instantly that we ought both to go and see what could be done in any way to relieve either the body or mind of the sufferer.

We said to each other while we were hastily dressing, "How shall we ever catch the horses? They have all been turned out, of course, as no one thought they would be wanted until Monday; and who knows where they have gone to?—miles away, perhaps; and it's pitch dark." Judge, then, of our delighted surprise, when, on going out into the verandah, preparatory to starting off to look for our steeds, we found them standing at the gate, ready saddled and bridled. It seemed like magic, but the good fairies in this case had been the two guests to whom I have alluded as having arrived just as we were starting for our picnic life. They were both "old chums," and understood the situation instantly. Whilst we were questioning Pepper (you can hear every word all over a New Zealand house), they had jumped up, huddled on their clothes, and gone over the brow of the hill to look for the horses. By great good fortune the whole mob was found quietly camping in the sheltered valley full of sweet grass, on its further side. To walk up to my pretty bay mare Helen, and lay hold of her mane, and then, vaulting on her back, ride the rest of the mob back into the stockyard, was, even in the deep darkness of a midsummer night, no difficult task for eyes so practised to catching horses under all circumstances. So here was one obstacle suddenly smoothed, and as I hastily collected my few simple remedies, consisting chiefly of flannel, chlorodyne, and brandy, I could only trust and pray that poor Fenwick's case might not be so desperate as Pepper represented it.

To our impatience, the difficult track, with its swamps and holes, its creeks to be jumped, and morasses to be avoided, seemed long indeed; but to judge from the continued profound darkness,—that inky blackness of the sky which is the immediate forerunner of daylight,—the dawn could not be far off. How well I remember the whole scene! F—— tied his white handkerchief on his arm, that Helen and I might have a faint speck of light by which to guide ourselves. Pepper rode close to me, pouring into my ears dismal predictions of Fenwick's end; whilst I, amid all my anxiety, could only think of the dangers of the track, and whether, in the pitchy darkness, we should ever get to the home station. The dew fell so heavily that more than once I thought it must be raining, but those were only wind-clouds brooding in the great dark vault above us. More welcome than ever sounded the bark of the dogs, which told us we had reached the end of our stumbling ride; and the moment their tongues woke up the silence, a lantern showed a ray of light to guide us to the hut door.

I jumped off my horse instantly, and went in. At first I thought my patient was dead, for he lay, rigid and grey, in his bunk. At a glance I perceived that nothing could really be done to help him whilst he was lying on a high shelf, almost out of my reach, in a small hut filled with bewildered men, who kept offering him from time to time a "pull" at a particularly good pipe, having previously poured all the grog they could muster down his throat, or rather over his pillow (his saddle performed that duty by night), for he had been unable to swallow for some hours. I remembered that there were the bedsteads we had used at the house, and also some firewood still left in the kitchen. Explaining to Pepper how he was to wrap poor Fenwick in every available blanket in the place, and carry him across the open space into the parlour, I hastily ran on before, got some one to help me to drag one of the light frames into the sitting-room, laced it before the fireplace, and then made up a good blazing fire on the open hearth. By the time the dry wood was crackling and sparkling out its cheery welcome, my patient arrived, and was laid down, blankets and all, on the rude little bedstead, before the blaze. By its fitful and uncertain light I proceeded to examine the enormous frame stretched so helplessly before me, feeling half afraid to touch him at all. F— — was very trying as an assistant, for he looked on without making any suggestions, and only said from time to time, "Take care: the man is dead." To my inexperienced eyes he indeed seemed past all human help. His skin was icy cold, and as wet as if he had been lying out in the dew. No flutter of pulse, nor sign of breath, could my trembling efforts discover; but I fancied there was the least little sign of pulsation about his heart. Of course I had not the vaguest notion of what was the matter with the man, for all Pepper could tell me was that "Fenwick's been powerful bad, you bet." This does not sound a minute diagnosis to go on, and the only remedies which presented themselves to my mind were those I had studied as being useful for the recovery of drowned persons. So to work I set, as if the poor fellow had just been fished out of the creek; and whenever any one wanted to teaze me afterwards they would declare I had insisted on Fenwick's being held up by his heels. But of course that was all nonsense. What I did really do was this, and a doctor in Christchurch, whom I afterwards consulted as to my treatment, assured me, laughingly, that it was "capital."

I made Pepper and another man both rub the cold clammy body, as hard as they could with mustard and hot flannel. I got some bottles filled with hot water (for it did not take five minutes to boil the kettle) and placed to his icy-cold feet and under his arms, then I mixed a little very strong and hot brandy and water, to which I added a few drops of chlorodyne, and gave him a teaspoonful every five minutes. For the first half-hour there was no sign of life to be detected, and the same horrible bluish pallor made poor Fenwick's really handsome face look ghastly in the flickering light. My two assistants were getting exhausted, and Pepper had more than once murmured, with the recollection of the past fortnight's work strong upon him, "Spell, oh!" or else "Shears!" whilst his companion inquired pathetically, "What was the use of flaying a dead man?" To these hints I paid no attention, though my damp riding habit was steaming from the heat of the fire and I felt dreadfully tired; for certainly there seemed to my eyes a healthier tinge stealing over the rigid features, and it could not be my fancy which detected a stronger effort to swallow the last spoonful of brandy.

I need not go into the details of my jumbled-up remedies; probably I should bring upon myself serious remonstrances from the Royal Humane Society, if my treatment of that unhappy man were made public. It is enough to say that I "exhibited" mustard by the pound and brandy by the quart, that I roasted him first on one side and then on the other, that his true skin was rubbed off, that I chlorodyned him until he slept for nearly a week, and that when he finally recovered he declared he felt "as if he'd been dead:" "And no wonder," as Pepper always remarked. The only clue I could get to the cause of his illness was a shy confession, about a week afterwards, that he had eaten a few mushrooms. Fenwick's idea of a few of anything was generally a liberal notion. I questioned him narrowly as to what he had had for supper the night he was taken ill, and this was his bill of fare:—

"Well, you see, mum, I wasn't rightly hungry: it must have been them gripses coming on. So I only had a shoulder (of mutton, bien entendu; when Fenwick had really a good appetite he regarded anything less than a whole leg of a sheep as an insult) that night, half-a-dozen slap jacks, and a trifle of mushrooms." "How big were the mushrooms?" I asked. "Oh, they was rather fine ones, mum, I won't deny: they might have been the bigness of a plate." Now even supposing them to have been perfectly wholesome, a few dozen mushrooms of that size, eaten half raw with a whole shoulder of mutton, are quite enough to my ignorant mind to account for so severe a fit of the "choleraics."

Chapter XVII: Odds and ends.

My nerves had hardly recovered the shock of having the care of such a huge patient thrust on me; for, seriously speaking, Fenwick took a good deal of nursing and attention before he got well again, when we had another night alarm. Our beautiful summer weather was breaking up; high nor'-westers had blown down the gorges for days, and now a cold wet gale was coming up in heavy banks of fleecy clouds from the sou'-west. Everything looked cold and wretched out of doors, but the sheep-farmers were thankful and pleased. Their "mobs" could find excellent shelter for themselves, for it takes very bad weather to hurt a Merino sheep, and the creeks had been running rather low. "We shall have a splendid autumn after this is over," said all the squatters gleefully, "with lots of feed: there's Tyler's creek coming down beautifully."

So I was fain to be content, though my fowls looked draggled and wretched, and my pet patch of mignonette became a miniature desert, its fragrance being all blown and rain-beaten away. Good fires of lignite and wood made the house cheery, and we went to bed, hoping for fine weather next day. In the middle of the night everyone was awakened by a tremendous, echoing noise outside, whilst the frail wooden house vibrated perceptibly. It could not be caused by the wind: for, although the rain kept pouring steadily down, the furious sou'-west gusts had long ago been beaten into a sullen silence by the descending torrents. For a moment, and half-awake, an old tropical reminiscence floated through my sleepy, startled mind: "Can it be an earthquake?" I dreamily wondered. But, no earthquake of my acquaintance was ever yet so resounding and noisy, for all its crumbling horror: yet, the house was certainly shaking. "What is it? What are you doing?" rang in shouts through the little dwelling, as its dwellers came thronging, one after another, to our door. Frightened as I was, I can perfectly remember how indignant I felt, when it became clear to my mind that they all thought we were making such an uproar. How could we do it, if even we had wished to get out of our warm beds, and create a disturbance on such a wild night.

"Good gracious! the house is coming down," I cried, as a fresh shudder ran through the slight framework of, our little wooden home. "Pray go out, and see what is the matter." Thus urged, F— — opened a casement on the sheltered side,—if any side could be said to be sheltered in such weather,—and cautiously put his head out. I peered over his shoulder, and never can I forget the ridiculous sight which met our eyes. There, dripping and forlorn, huddled together under the wide roof of our summer parlour, as the verandah used to be often called, the whole mob of horses had gathered themselves. The garden gate chanced to have been left open, and, evidently under old Jack's' guidance, they had all walked into the verandah, wandered disconsolately up and down its boarded floor, and after partaking of a slight refreshment in the shape of my best creepers, had proceeded to make themselves at home by rubbing their wet sides against the pillars and the wooden sides of the house itself.

No wonder the noise had aroused us all. Ironshod hoofs clattering up and down a boarded verandah is riot a silent performance; and Jack was so cool and impudent about it, positively refusing to stir from the sheltered corner by the silver-pheasants' aviary, which he had chosen for himself. The other horses evidently felt they were intruders, and were glad enough, on the flapping of a handkerchief, to hurry out of their impromptu stables, making the best of their way through the narrow garden gate, and so out upon the bleak hills again. But Jack's conduct was very trying; he found himself perfectly comfortable, and evidently intended to remain so; neither for wishing nor coaxing, for fair words nor foul, would he stir. It seemed so horrid to have to dress and go out in such a downpour of rain, that we weakly deliberated on the expediency of letting the cunning old stock-horse remain; but fortunately, at that moment he began to scratch his ear with his hind foot, waking up a thousand echoes against the side of the house as he did so, and making the pictures dance again on the canvas and paper walls. "This will never do," cried we all, desperately: "he sure must be taken to the stable or he'll come back again." That was exactly what Jack meant and wanted: so to the stable he went, under poor shivering Mr. U— —'s guidance, and the old rogue spent a dry, warm night under its roof.

It was the more absurd Jack pretending to be afraid of a wet night, when he had walked many and many a weary mile over the rough mountain passes towards the West-Coast, with a heavy pack on his back and in all sorts of weather. A tradition existed in our neighbourhood that Jack had once been met crossing the Amuri Downs with a small barrel-organ, an American cooking stove, and a sow with a litter of young ones, all packed on his back, "and stepping out bravely under them all," as my informant added. But I cannot vouch for the truth of the items of this load. Jack's fame as a stock-horse, as well as a pack-horse, stood high in the Malvern Hills, but his conduct in the shafts was eccentric, to say the least of it. He could not bear to be guided by his driver, and was always squinting over his blinkers in the most ridiculous manner. If he perceived a mob of cattle or horses on a distant flat, he would set off to have a look at them and determine whether they were strangers or friends, dragging the gig after him "over bank, bush, and scaur."

Once when we were in great despair for a cart-horse, Jack was elected to the post, but long before we had come to the journey's end we regretted our choice. It was during the first summer of my life in the Malvern Hills, and whilst the nor'-westers were still steadily setting their breezy faces against such a new fangled idea as a lawn. I had wearied of sowing grass seed at, a guinea a bag, long before those extremely rude zephyrs got tired of blowing it all out of the ground. There was my beautiful set of croquet, fresh from Jacques, lying idle in its box in the verandah, and there was my charming friend, Alice S— —, longing for a game of croquet. When pretty young ladies wish for anything very much, and the house is full of gentlemen, it goes hard, but that they get the desire of their innocent hearts. So it was in this case. One fine afternoon Alice wandered into the verandah and peeped for the hundredth time into the box. "What beautiful things," she sighed, "and how hard it is we can't have a game." "I know a patch of self-sown grass," sang one of the party, "whereon we might play a game." "Where: oh, where?" we asked, in eager chorus. "About two miles from this, near a deserted shepherd's hut; it is as thick and soft as green velvet, and the sheep keep it quite short." "Is the ground level?" we inquired. "As flat as this table," was the satisfactory answer.

Of course we wanted to start immediately, but how were we to get the croquet things there, to say nothing of the delightful excuse for tea out of doors which immediately presented itself to my ever-thirsty mind. A dray was suggested (carriages we had none; there being no roads for them if we had possessed such vehicles); but alas, and alas! the proper dray and driver and horse were all away, on an expedition up a distant gulley getting out some brush-wood for fires. "There's Jack," some one said, doubtfully. He had never even drawn a dray in his life, so far as we knew, but at the same time we felt sure that when once Jack understood what was required of him, he would do his best to help us to get to our croquet ground. So we flew off to our different duties. Alice to see that the balls, hoops, and mallets were all right in numbers and colours, &c.; I to pack a large open basket with the materials for my favourite form of dissipation—an out-door tea; and the gentlemen to catch Jack and harness him into the cart.

Peals of laughter announced the setting forth of the expedition; and no wonder! Inside the dray, which was a very light and crazy old affair, was seated Alice on an empty flour-sack; by her side I crouched on an old sugar bag, one of my arms keeping tight hold of my beloved tea-basket with its jingling contents, whilst the other was desperately clutching at the side of the dray. On a board across the front three gentlemen were perched, each wanting to drive, exactly like so many small children in a goat carriage, and like them, one holding the reins, the other the whip, and the third giving good advice. In the shafts stood poor shaggy old Jack, looking over his blinkers as much as to say, "What do you want me to do now?" Our good humoured and stalwart cadet Mr. U— —, walked backwards, holding out a carrot and calling Jack to come and eat it.

In this extraordinary fashion we proceeded down the flat for two or three hundred yards, one carrot succeeding the other in Jack's jaws rapidly. Mr. U— — was just beginning to say "Look here: don't you think we ought to take turns at this?" when Jack caught sight of a creek right before him. He only knew of one way of crossing such obstacles, and that was to jump them. No one calculated on the sudden rush and high bound into the air with which he triumphantly cleared the water; knocking Mr. U— — over, and scattering his three drivers like summer leaves on the track. As for Alice and me, the inside passengers, we found the sensation of jumping a creek in a dray most unpleasant. All the croquet balls leapt wildly up into the air to fall like a wooden hailstorm around us. The mallets and hoops bruised us from our head to our feet; and the contents of my basket were utterly ruined. Not only had my tea-cups and saucers come together in one grand smash, but the kettle broke the bottle of cream, which in its turn absorbed all the sugar. Jack looked coolly round at us with an air of mild satisfaction, as if he thought he had done something very clever, whilst our shrieks were rending the air.

What a merry, light-hearted time of one's life was that! We all had to work hard, and our amusements were so simple and Arcadian that I often wonder if they really did amuse us so much as we thought they did at the moment. Let all New Zealanders who doubt this, look into those perhaps closed chapters of their lives, and as memory turns over the leaves one by one, and pictures like the sketches I try to reproduce in pen and ink, grow into distinctness out of the dim past, it will indeed "surprise me very much," if they do not say, as I do,—my pleasant task ended,—"Ah, those were happy days indeed!" ended,—"Ah, those were happy days indeed!"